Get Your Goat

How to Keep
Happy, Healthy Goats
in Your Backyard,
Wherever You Live

Get Your Goat

Brent Zimmerman

Quarry Books
100 Cummings Center, Suite 406L
Beverly, MA 01915
quarrybooks.com • craftside.typepad.com

First published in the United States of America in 2012 by
Quarry Books, a member of
Quayside Publishing Group
100 Cummings Center
Suite 406-L
Beverly, Massachusetts 01915-6101
Telephone: (978) 282-9590
Fax: (978) 283-2742
www.quarrybooks.com

10 9 8 7 6 5 4 3 2 1

ISBN: 978-1-59253-757-0

Digital edition published in 2012
eISBN: 978-1-61058-1-875

Library of Congress Cataloging-in-Publication Data
Zimmerman, Brent.
Get your goat : how to keep happy, healthy goats in your backyard, wherever you live / Brent Zimmerman.
p. cm.
Includes index.
ISBN-13: 978-1-59253-757-0
ISBN-10: 1-59253-757-X
1. Goats. I. Title. II. Title: How to keep happy, healthy goats in your backyard, wherever you live.
SF383.Z55 2012
636.3'9--dc23

2011031321

Get Your Goat contains a variety of tips and suggestions for keeping goats. While caution was taken to give safe recommendations, it is impossible to predict the outcome of each suggestion. Neither Brent Zimmerman, nor the publisher, Quayside Publishing Group, accepts liability for any mental, financial, or physical harm that arises from following the advice or techniques, using the procedures in this book. Readers should use personal judgment when applying the recommendations of this text.

Design: Kathie Alexander
Developmental editor: April White
Cover images: top row, istockphoto.com; main image, shutterstock.com; spine, fotolia.com; back cover, top, ©Teri Myers/braidedbowerfarm.com;
back cover, bottom, © Peter M. Bergin/braidedbowerfarm.com
illustrations: Judy Love

This book is dedicated
to the many new friends
I otherwise would not have met
had it not been for the goats.

CONTENTS

CHAPTER 1

Introduction

"Goats, huh?" your neighbors might say, with a wide grin or a nervous laugh and a glance at their prized rosebushes, when you tell them of your plans for a backyard barnyard.

You aren't alone in your daydreams of living a simple life. You can go back to the land even if you don't have much land to go back to. Urban farmsteads are popping up all over. From keeping a compost pile full of earthworms and a small chicken coop to a beehive on the rooftop and goats in the garage, a growing number of your neighbors are committing to growing and sourcing their own food.

I credit the farmers' markets. It's inspiring to visit your local market and see the hardworking farmers and the produce, meats, and cheese they bring every week. You may find yourself saying, "I would love to do this, plant that, make those," as you go from stand to stand. Maybe you find yourself drawn to the goat cheese sellers, daydreaming of your own herd as you look at the photo album of their goats on display and sample just one more cheese.

This book shows you how to turn your backyard barnyard daydreams into a reality, introducing you to goats, their needs, and the responsibilities of the goat farmer. It addresses issues particular to the backyard farm—leash training your goat for a walk around the block—and topics that apply to goat farmers both urban and rural, including breeding and milking. This guidance comes from a lifetime of raising animals and twenty years (and counting) of goat-keeping experience ranging from a few goats in the garden shed to a large commercial dairy of eighty milking does.

There are many places around the world where it's quite common to see goats walking up and down the streets unattended—but that's probably not true in your neighborhood. Still, attitudes and laws regarding goats are changing. As urbanites rediscover the usefulness and friendliness of these productive, happy animals and the health benefits that goat milk, cheese, and meat provide, towns and cities are slowly but surely opening themselves up to change.

City Goat, Illegal Goat?

Before you start digging fence-post holes, you must find out if keeping goats is allowed in your town or city. Many places have laws on the books about keeping livestock. These laws may limit the numbers of goats allowed, dictate the size of the lot required for keeping goats, or prohibit you from keeping a buck. Or the laws may outlaw goats completely.

To find out what laws govern livestock in your area, contact the town or city government for the applicable zoning ordinances.

If goats are outlawed, don't give up. As more people embrace the idea of urban farmsteads, towns and cities are opening up to the idea of keeping small numbers of animals. Ask what you can do to have the law changed.

Backyard farmer Jennie Grant did just that. Grant wanted to keep two goats at her home in Seattle, Washington, but city laws classified goats as farm animals and prohibited farm animals in the city. Grant wondered, "If people can have a Rottweiler, why can't I have a mini goat?"

Grant lobbied the Seattle City Council, and the law was changed. Thanks to Grant, Seattle residents can now keep two hornless goats, up to 100 pounds (45.4 kg) each, in city backyards.

Grant's success has inspired others, and her Goat Justice League dispenses advice to would-be goat farmers who face city or town restrictions.

Tip

The Goat Justice League of Seattle is just one organization that offers classes and workshops on goat keeping. Many other cities and locales have online communities that support your new hobby from many angles. See Resources (page 151) for more information.

A herd of Saanens enjoy their breakfast in a well-planned, easy-to-manage feeding set-up.

Why We Farm

There are many reasons to think about starting a backyard goat farm: Maybe you want to reconnect with the food chain by participating in the daily milking chores. Maybe you have extra space in your lot that would be better suited to goats than gardening. Maybe you have children who would like to participate in the local 4-H or other agricultural club. Or maybe you want to turn unsightly weeds into food.

At its core, the practice of farming animals has always been about turning inedible plants (for humans) into an edible food, and goats are about as good at that game as any animal. It is amazing to think that milk and creamy white goat cheese started out as pine needles, tree bark, and fallen leaves. The goat chewed and swallowed, brought it back up to be chewed again, and swallowed it again. It was mixed with water and saliva, digested into the bloodstream, passed through the circulatory system several hundred times, secreted into the milk ducts, and coaxed into the pail by the farmer.

A room (plank) with a view: This simple raised ramp will provide years of enjoyment and exercise for your goats.

Why Goats?

They are smaller than a large dog, and some say smarter, and they make milk for cheese! What's not to like about goats? Goats thrive on attention but seem just as content to be left alone, leaving you time to do other things around the farmstead. Goats adapt very quickly to many living situations, so your backyard could be goat paradise.

Goats can provide you with a variety of products such as milk, butter, cheese, manure for the garden, and companionship or meat, hides, and fiber.

Milk and cheese: A fresh glass of milk or a bite of cheese from your very own goat will taste better than anything you can find in the store. They will also cost you more—in time dedicated to twice-daily milkings for six to ten months—than if you bought it from the store, but the rhythm of daily chores is a bonus for some.

Worldwide, goat's milk is more popular than cow's milk. It may be healthier, too. Goat's milk is more similar to a human mother's milk than cow's milk, and many people who have lactose intolerance problems with cow's milk are able to drink goat's milk trouble free.

Fiber: Fiber goats are great pets and provide a relaxing hobby. They are perfect for spinners and fiber artists who want to generate their own raw materials. In limited numbers of goats, it's not too much work to shear, clean, sort, card, and spin the fiber.

Meat: If you want your own meat source, goat kids give a good-size carcass from about the age of six weeks. At approximately nine weeks they will have a carcass weight (meat, edible interior organs, bones) of about 32 to 35 pounds (14.5 to 15.9 kg) depending on the breed and management.

Pets: Goats are smart and happy and like attention and affection. They are also very independent as far as pets go. They must also be penned, watched carefully if out, or put on a leash so they don't destroy any plant within reach. All backyard goats are pets to some degree, in terms of attention, care, and proximity.

The Backyard Barnyard

Keeping goats in small spaces and city settings is nothing new. It's been done for centuries. Historically it was more out of necessity than hobby. (Though who's to say your hobby isn't a necessity?) And goats can thrive in an urban setting if you just have a little ingenuity, a lot of common sense, and total commitment to being the caretaker of these wonderful animals that give so much while asking for so little in return.

Of course, keeping goats isn't all sunshine and half-eaten rosebushes. There is work involved, and the job of the backyard farmer includes keeping not only yourself, but your goats, your neighbors, and possibly, the town happy. You have to be a responsible owner and make sure your animals are well cared for and not a nuisance to others. If through improper management you allow your goats to become smelly or loud or to wander about tasting all the flowers on the block, it's only natural for neighbors and zoning officials to take a second look at these backyard farms, causing problems not only for you but for all urban farmers. So remember, you are representing not only yourself but backyard goat farmers everywhere.

No matter where you live, the key to raising goats successfully is having a passion for what you are doing. As with any new project or idea, goat keeping is only difficult if you don't enjoy it. When you are invested in your goats' well-being and their happiness, in making great cheese and enjoying that perfect fresh milk every morning, the challenges that come with backyard farming become joyful, not dreadful. Goats will try your nerves, try your patience, and will even try your daffodils. Having a real passion for your animals, your food, and where your food comes from will have you smiling even as you say through clenched teeth, "Oh well, that's goats for ya!"

This little pygmy goat doesn't seem to be bothered by what kind of mountain he climbs as long as he is king.

Goats will try your nerves, try your patience, and will even try your daffodils.

From the Farm: The Making of a Farmer

It was destiny that I found myself on a hillside farm surrounded by animals. When I was a kid, my brothers had an electric race-car track. I played with a plastic farm set. Now that we're grown, my brothers race real cars and I milk goats every day.

When I was ten years old, my friend D. W. and I decided to go into the dairy goat business. We wrote out an elaborate business plan without even knowing what a business plan was. Looking back, it was actually a pretty good plan, complete with cost and profit projections and how we would cover those costs. My grandpa owned the farm in Michigan where I grew up, so I made my presentation to him. It was simple: two goats and I do all the work.

It was a different time, and my grandpa was a man of few words. Nervously I stood in front of him and read my plan. When I had finished, he looked me straight in the eye and said, "No." And that was that.

It would be another sixteen years before I got my first goat, but I've made up for lost time. I'm now the co-owner and full-time milker and cheese maker of a beautiful Tuscan paradise in the hills of Italy.

Italy wasn't part of the plan when I was playing with that plastic farm set. But when I was in my mid-twenties, I came to Italy on what was supposed to be a few weeks' vacation. At that time, in these hills, there were still flocks of sheep dotting the landscape followed by ladies wearing scarves on their heads. Small herds of cattle roamed the hilltops unattended, and homemade cheese was easy to find.

The author sneaks in a quick snuggle before serving breakfast milk to his kids.

I moved to a farm, received a goat as a gift, bought a flock of sheep, milked my cows, and made my cheese. I learned farming, cheese making, Italian, and patience all at the same time. I became as self-sufficient as I could, sowing fields by hand, grinding my own grains, baking bread, butchering my animals, tending my garden, and preserving the harvest. My favorite books then and now are "how-to" guides, usually about a farm animal—books a lot like the one you have in your hands right now.

Those books—and this one—will tell you that the life of a farmer isn't always easy. Mine hasn't been, but the animals keep me here. As I write this, I have fifty purebred Oberhasli goats in the barn. It is a joy to go to the barn every morning, and I am never happier than when I take my goats out for a walk through the woods and pastures surrounding the farm. They all perk up their ears at once when I shout out "let's go" as we move on to the next grazing area. I am proud of the connection I have with my animals and this farm.

The hook to this lifestyle I love so much is that you can't leave it. Every season has something to offer, and as one thing matures and withers, something new comes to life. As much as I look forward to the long summer days while shivering in front of the woodstove, I look forward to those dark nights of winter—the season of sleep, reflection, and planning—each busy summer.

The experiences of more than two decades on the farm have shaped my very simple approach to raising goats: I put myself in their hooves every day. I take a look at their diet, their sleeping area, and their exercise options. I consider their safety and their comfort. I ask myself, would I want to be a goat here? Is there something that needs to be improved to help the goats live happy, stress-free, and therefore productive lives?

I provide for them, and they in return give me large quantities of high-quality milk. For the past nine years, I have had a licensed cheese dairy transforming every drop of their milk into delicious cheese that I sell at farmers' markets throughout Tuscany. It's true: happy goats really do give happy milk.

I've had my share of success as a farmer, and I have been humbled on more occasions than I care to remember. All of it has taught me one thing: give your best and your goats will give their best.

Every season has something to offer, and as one thing matures and withers, something new comes to life.

A pair of blue-eyed does and their multicolored kids show their inquisitive nature.

CHAPTER 2

The Goat

It's important to start your backyard goat-farming project on the right hoof. The first step to a successful backyard goat farm is educating yourself about the goat itself.

The domestic goat of today *Capra aegagrus hircus*, is a subspecies of *Capra aegagrus*, the wild goat. Evidence suggests that goats first lived in cooperation with man over 7000 years ago. They're sometimes compared to and quite often found living together with sheep. Not so long ago it was a common sight to see a flock of sheep with a few goats mixed in. Although similar in size and both being ruminants, there are some easy-to-note differences. Sheep are grazers and goats are browsers, meaning sheep prefer grass and goats prefer eating bushes, trees, vines, and brambles. Like their relatives the sheep, they are even toed, ungulates, cud chewers, and have triple purposes, providing meat, milk, and fiber.

Understanding these cute, curious creatures at a basic level—their anatomy, their life cycle, their instincts—is key to understanding how to fulfill their needs throughout their lives. This introduction to goats is not a comprehensive, zoological study but a targeted lesson on what you need to know to start raising them in your backyard. It is also an introduction to goat farmers, their responsibilities, and their rewards.

Goat Keeper's Glossary

Part of the fun with any new undertaking is learning a new language. This chapter includes the most important basic terms for talking about your goats. Look for additional "Goat Keeper's Glossary" definitions in later chapters.

Buck: A male goat able to mate and sire offspring

Buckling: A young male goat, so called until he starts mating, usually not before eight months

Doe: A female goat of producing age

Doeling: A young female goat, so called until the she gives birth for the first time, usually twelve months

Dry doe: A doe that is not currently producing milk, as during the two months prior to giving birth

Heat: When does are ovulating, they are said to be in heat

Herd: Two or more goats living together

In milk: A doe that is currently producing milk daily

Kid: A baby goat, male or female, at least up until weaning of about eight weeks

Kidding: This is the act of having a kid. A doe giving birth is said to be kidding.

Milker: A doe who is in milk

Milking: The act of taking milk from a doe by hand or by machine

Producing age: The age during which a female goat is able to give birth to kids and produce milk, usually year one through ten

Rut: Bucks go into rut when the does start going into heat. When in rut, they have one thing on their mind. They will give off a strong odor and possibly act more aggressive. They're basically showing off for the doe in hopes of mating.

Wether: A male goat that has had his testicles removed by castration

Basic Goat Anatomy

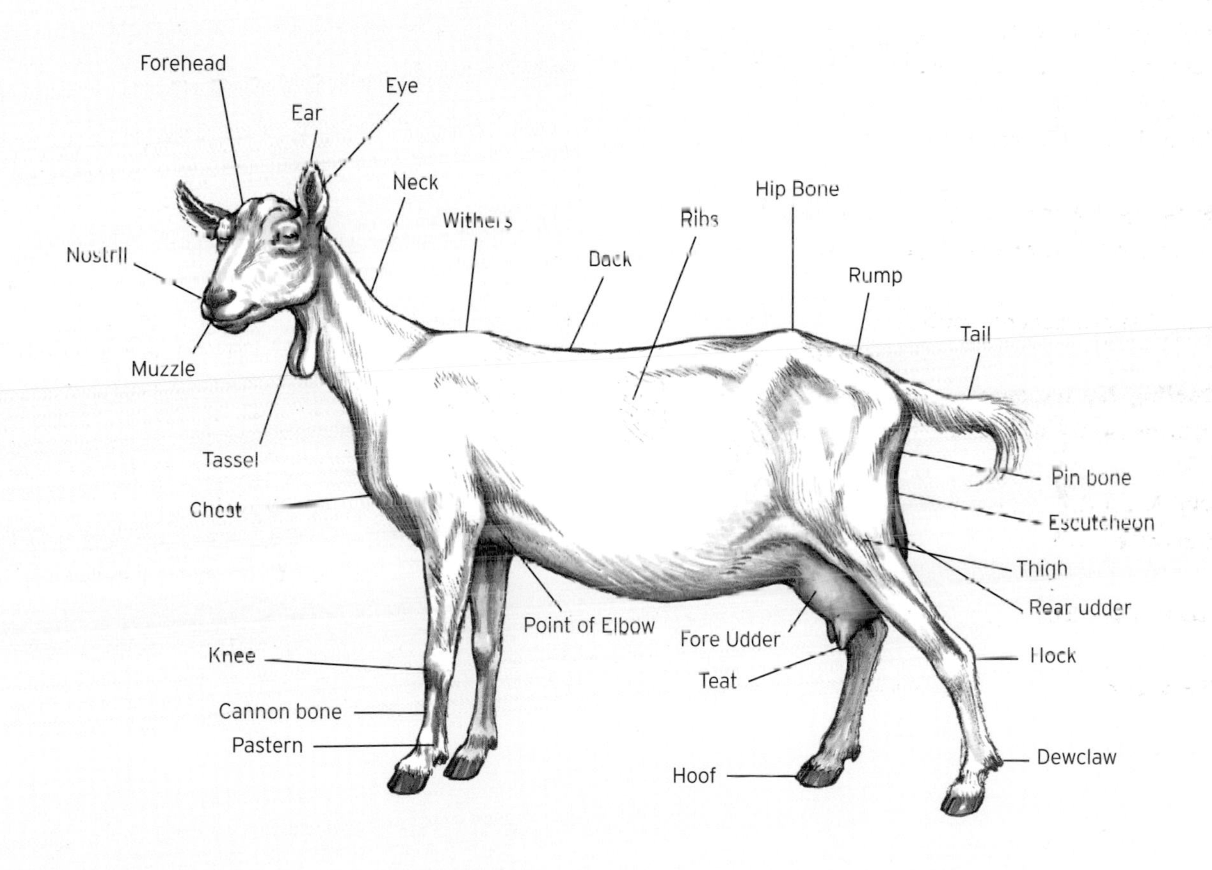

The Birds and the Bees of Goats

Bucks are handsome, proud, energetic, powerful animals. And they are necessary. Without the buck, we would not have kids or milk. But keeping a buck is often a poor choice for the backyard goat farmer—and not just because of the odor he produces when he is in rut.

Bucklings can reach sexually mature at two months and must be housed separately from doelings after the age of seven weeks to avoid unwanted pregnancies among doelings too young to handle pregnancy and kidding.

If he will be used as a herd sire, he will spend his first seven months playing, head butting, mounting his stablemates, learning the skills he will need when he starts mating at eight months. Bucks should be used for breeding sparingly their first year.

As the buck grows large and formidable, he will need ample food, plenty of exercise, and adequate space and strong fencing to keep him separated from the females. A strong buck separated from does that he can see and smell is difficult to keep fenced in. (Learn more about fencing in chapter 4.)

Bucks should be approached with caution at all times. A buck's idea of play—head butting and charging—can cause serious injury, especially to children who are at eye level with the buck.

Simply put, bucks can be more trouble than they are worth for a backyard goat farmer who can easily arrange to breed his or her does with a buck at a nearby farm. (Learn more about breeding in chapter 8.)

A Goat's Life

You may hear about does who live fifteen years, still giving three liters of milk a day, but this is the exception, not the rule. There are many factors which can add or detract years from a goat's life, such as proper feeding, protection from the elements, and responsible management and breeding.

Does on average live ten to twelve years, although most will only produce kids (and milk) for eight to ten years. The sad fact of life is that most does die as a result of complications during birthing. Bucks can live eight to ten years. Spending too much energy on rut can shorten their lives. Wethers usually live the longest, twelve to fifteen years. Wethers don't have the stress of rut, kidding, and giving up milk, so their lives are carefree and long lived.

After the first, each year of a goat's life—and each year of a goat farmer's life—follows a similar cycle, depending on the animal's use. The monthly guide outlined here roughly corresponds to the seasons of the Northern Hemisphere. For the Southern Hemisphere, January becomes July, February becomes August, and so on.

A Year in the Life

January

For the goat: This is a season of rest. The does are growing larger with pregnancy and seem content to lay about all day chewing cud, which is one way that goats produce heat to ward off the chill of deep winter.

For the goat farmer: This is also a season of rest. This is the time to plan for the kids that will arrive next month.

February

For the goat: In early February, the does are nearing the end of pregnancy. They start choosing their preferred lounging areas, digging a nest as their mothering instincts become stronger. In the second half of February, the kids start to arrive.

For the goat farmer: Early February is a time of preparation, stocking supplies, and arranging a vet for the coming kids. Late February is exciting. After five months of waiting, the kids are here. A goat farmer doesn't get much sleep in late February and early March.

March

For the goats: There may be more babies to be born, and the kids born in February are growing rapidly. The does stay close to their kids and begin producing more milk than the kids can consume.

For the goat farmer: With the arrival of the does' milk, comes a twice-a-day milking schedule and the start of cheese production.

April

For the goats: By April, most does have given birth, and the does and kids have settled into a happy mother/kid routine.

For the goat farmer: Twice daily milking and cheese production continues. April is also the time to sell any kids you don't plan to keep.

January

February

March

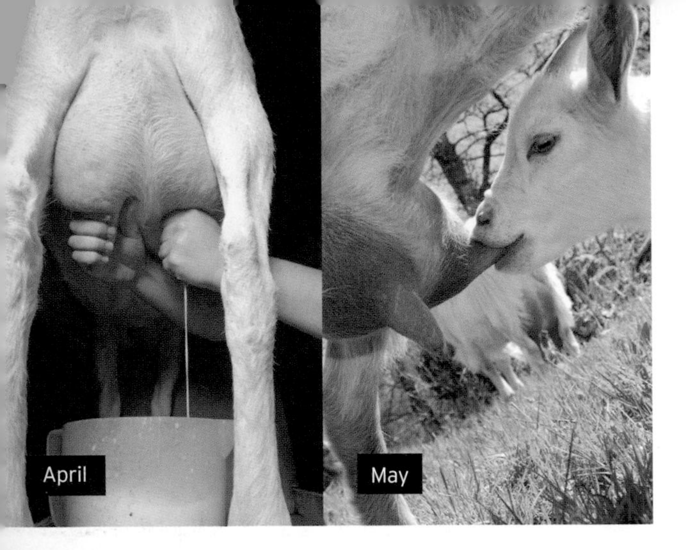
April

May

June

August

May

For the goats: The kids are weaned from their mothers' milk. If you are raising fiber goats, now's the time for shearing. The weaned kids will need to find a new home. Are you growing your herd? If you are home butchering, you can begin anytime.

For the goat farmer: Twice-daily milking and cheese production continues. The end of May is also haymaking time. Whether cutting your own hay or buying from another source, the best time for hay buying is during hay season, as prices will be much less than if you wait until mid-winter.

June

For the goats: As the temperature starts to climb, the goats will be less active during the hot afternoons, seeking out shade and needing plenty of fresh drinking water. By late June, bucks are separated from does, as breeding season is right around the corner.

For the goat farmer: Milking, cheesing, haying is the mantra.

July

For the goats: Does continues to produce milk, although the heat of summer may diminish production somewhat. They are also preparing for breeding season and need lots of healthy food to replace the nutrients lost through milking. Bucks begin to go into rut.

For the goat farmer: There's more milking and more cheese making. If you aren't keeping a buck, make sure you have your buck lined up by calling a breeder and making a breeding agreement so when your does are ready, you can go at a moment's notice.

August

For the goats: Both does and bucks need good nutrition in preparation for breeding season to ensure healthy kids and parents.

For the goat farmer: Milking and cheese making continue, plus necessary preparations for breeding season.

September

For the goats: This is breeding season. The five-month gestation period means kids will be born in February and March.

For the goat farmer: Milking and cheese making continue. It's time to carefully observe your docs for changes in behavior as they go into heat—and to get used to the smell of a buck in rut, if you keep one.

October

For the goats: After the does become pregnant, their milk will start to taper off to the point that you can stop milking. Shearing for fiber goats should take place.

For the goat farmer: Milking and cheese making slows down.

November

For the goats: The does are dry and progressing in their pregnancies. The goats are as happy as the goat farmer to enjoy the break from twice-daily milkings. Decide with your vet when to check your goats for worms and yearly shots for your area. November is a good time, as milking is finished and babies are still a few months away.

For the goat farmer: With relief and sadness, milking and cheesing finally comes to an end for the season.

December

For the goats: The year comes full circle, and the goats, no longer milking and starting to plump up with pregnancy are happy to rest.

For the goat farmers: The farmer rests, too.

October

November

December

The First Year

Kids are usually born in late winter or early spring. After eight weeks of nursing, either from their mother or from a bottle, they should be eating enough hay and feed to be safely take them off the nipple.

Doelings are capable of becoming pregnant as early as seven weeks of age but should not be bred until they are seven or eight months of age. Doelings usually have their first kid at or after twelve months. Some goat owners like to wait until a doeling is eighteen months old before breeding to allow her to grow bigger and stronger before being mated. The downside of this is that you lose one year of milking and kids. Breeding at seven or eight months means a doe joins the milking group at twelve to thirteen months of age to start her career in the dairy and cheese industry. Breeding at eighteen months means waiting until she is at least two years old before getting your first drop of milk.

Goat Psychology 101

To raise goats successfully, we need to understand them. Observation is your best ally, so you will be better able to give your goats what they need. The first thing you'll notice when observing your goats is that goats are herd animals and need companionship.

Goats are prey animals and thousands of years of being hunted by predators have etched into the DNA of today's goats the idea that there is safety in numbers. That's why goats are known for being excellent stable mates for horses and for bonding with children or even dogs. But the easiest and most natural companion for a goat is another goat.

Many a new goat keeper will excitedly visit a goat barn and bask in the tranquil setting of quiet cud chewing, attentive mothers, and frolicking kids. After taking a happy, bouncing young goat home, however, they watch her wither into a sad, lost-looking creature, constantly bleating, possibly going off food and growing weak or worse, all because she's lonely. Or a lonely goat may be anxious and repeatedly try to escape her enclosure.

The goat keeper's responsibility to his or her goats is to give them the most stress-free, natural life possible so the goat can do what she does best: produce kids and milk. Therefore, it is recommended that you always keep at least two goats.

Even in a herd of two, goats establish a hierarchy. Unless they are raised together, your goats may duel it out with repeated head butts, locking of horns, and other bullying. Try as you might to stop it, your goats will keep at it until there is a clear winner.

In your small backyard herd, you will clearly see which goat is dominant and which is submissive. The submissive animals will move over when a dominant goat comes to the feeder, and the leader will sometimes intimidate the weaker goat for her sleep space. You may find yourself scolding, "Would you leave her alone, please?" but remember that it is a natural process and still, your goats are happier together than alone.

The goat keeper's responsibility to his or her goats is to give them the most stress-free, natural life possible so the goat can do what she does best: produce kids and milk.

A shy, young Saanen goat finds comfort with a good friend.

CHAPTER CHECKLIST

- ☐ Have you learned the important goat terminology?
- ☐ Have you considered the life of the goat and the responsibilities of the goat farmer, year-round?
- ☐ Do you understand why goats need a herd?

From the Farm

Lessons from Babbi

I used to help an elderly lady and her very elderly mother with their small farm, cutting wood, making hay, and cleaning out the barn. They had a large flock of sheep and a few Saanen goats for milk.

It was my first summer in Italy, and I was dreaming of filling my farm with animals. When the neighbors' goats kidded, they gave me a beautiful little doeling as a thank you. I named her Babbi.

Babbi spent her early days following me everywhere, and I loved the company of this happy little goat sticking her nose in everything I did. A year passed like this, and the house and the farm slowly started to take shape. The first few times I found little goat prints in the freshly laid cement, it was charming. But soon the damage Babbi caused started to get annoying.

I needed to renovate a house, not babysit a goat. I put Babbi in a fenced pen, where she wouldn't cause me stress, but she screamed and bleated until she was hoarse. She stalked the fence line. She got her head stuck and I would spend half my day getting her horns untangled. I tethered her where she could be near me but out of the way, but she walked in circles until the tether was tight against the tree.

What I didn't know as a first-time goat farmer is that a goat needs a herd. I was Babbi's herd. Because I couldn't dedicate all my time to her anymore, she was lonely and I was spending all my time calming her. I was not building a house.

The story has a happy ending: Babbi moved on to a long, happy life with an elderly gentleman who kept a small herd of very pampered goats, and I decided that was the end of my life as a goat farmer. Two years later, though, armed with the knowledge of what a goat needs, I brought home two kids. Twenty years later, I have a barn full of forty-five milking does and seventy-five babies.

CHAPTER 3

Goat Breeds

Just as for everything there is a season, for every goat there is a reason. We keep goats because they provide us with something, be it milk, meat, fiber, ecological weed control, or simply the joy of their company.

A breed can be defined as genetically related animals that share specific characteristics. When you breed (mate) two goats of the same breed (same characteristics), the resulting offspring will have the same breed characteristics as the parents—color, body type, productive abilities, adaptability, strengths, and weaknesses.

There are more than three hundred goat breeds in the world. They range in size as small as 15 inches (38.1 cm) to more than 40 inches (1 m) tall, weighing 50 pounds (22.7 kg) to more than 300 pounds (136 kg). They come in a variety of colors—white, black, brown, reds, gray, fawn, blonde, and any mix of the above. Some of these breeds evolved in the wild, through a breed's isolation and survival abilities; others developed through human intervention—selective breeding—to produce certain desired traits, such as a specific size or color, or to enhance productive capacity such as meat, milk, or fiber.

As farmers, large scale or backyard, our goal is to be sustainable, which means we need to maintain healthy goats that produce the most usable products in the most efficient way we can. Goats categorized as dairy breeds produce larger amounts of milk for the same amount of input than those categorized as meat breeds, and meat breeds put more weight on the hoof than dairy breeds for the same amount of input.

Pygmy goat

A Boer goat with two kids

Choosing a Breed

Once you decide what you want your goats to provide you—milk, meat, fiber, or companionship—choosing a breed is really a matter of personal choice, considering your needs. Do you prefer the look of one breed over another? Does a smaller breed make more sense than one of the larger breeds? Whether you are considering milk, meat, or fiber breeds, your decision will be based on what personality and physical traits you like.

Goats have personalities that are by nature friendly and curious. Various breeds may have slight differences, but within the sometimes bossy Alpine breed you find big sweethearts as calm as kittens and among the most docile of the Nigerians you are sure to come across a real tough bird. Ultimately, when given suitable housing, excellent care, and consistent, respectful handling, all goat breeds will prove to be enjoyable and big hearted.

How to Choose: Milk, Meat, or Fiber?

Before you choose a particular breed of goat for its beautiful markings or sweet temperament, consider what your goal is. Do you want your goat to produce milk, meat, or fiber?

There is an informal fourth category: pet. The most important consideration in choosing a pet goat is personality, but remember: If you are only looking for backyard companionship, you probably won't want to feed a large Boer or milk a gallon (3.8 L) a day from a Saanen.

Shearing an Angora goat

Before choosing a dairy breed, ask yourself the following:

- Do I want all the milk the goat may produce (as much as 343 gallon [1,300 L] a year)?
- Do I like the taste of a particular breed's milk over another?
- If I am going to make cheese, will the goat produce milk with the butterfat percentage that I need for great cheese?

Before choosing a meat goat, ask yourself the following:

- Do I have the space required to house the goat not only as a young kid but also as a full-grown adult?
- Will I have to buy all of the goat's food or will the goat be able to forage in my backyard?
- Do I have the means to transport an animal weighing 100 to 300 pounds (45 to 136 kg) for mating or butchering?
- Is my backyard suitable for home butchering or is there a butcher nearby willing to come to me or within driving distance if I don't want to do the job myself?

Before choosing a fiber goat, ask yourself the following:

- Does my region's climate suit the goat?
- Do I like the goat's coloring?

Dairy Breeds

Most backyard goat farmers raise goats for milk. The main attribute of the dairy goat is her ability to produce an elevated quantity of milk over a long period of time, usually eight to ten months each year.

When considering milk from a dairy goat, there are two things to consider: how much milk she produces and the percentage of butterfat in the milk.

Different dairy breeds produce different amounts of milk, from about 1,826 pounds (or 830 L) to about 2,350 pounds (1,185 L). In some markets, including the United States, milk is commonly measured by weight because fresh milk foams, making it hard to measure in volume. Throughout this book, however, we will measure milk in liters as it's easier to understand how much milk we are talking about. (One liter is slightly more than one quart.)

"Percentage of butterfat" is how people in the dairy industry measure the fat content and the cheese-making potential of milk. The higher the percentage of butterfat, the more butter or cheese you will be able to make per pound (0.45 kg) of milk. A general rule of thumb is the less milk a breed gives, the higher percentage of butterfat the milk contains and the more milk a breed gives, the lower the percentage of butterfat. The Swiss mountain breeds produce more milk per year than the Southern breeds, but the butterfat of the milk from the Swiss breeds is on average lower than the Southern breeds.

Tip

A general rule of thumb is the less milk a breed gives, the higher percentage of butterfat the milk contains and the more milk a breed gives, the lower the percentage of butterfat.

Alpine

Alpine

Also known as French Alpines, these goats are graceful animals with upright ears and medium-to-short hair. Alpines come in all colors, but pure white Alpines and Alpines with Toggenburg markings and color (light brown with white face markings) are not allowed in the breed registry because of their similarity to the Saanen and Toggenburg breeds.

Origin: Alpines originated in the western Alps in south-central Europe. In 1922, eighteen does and three bucks were imported to the United States. All full-blooded Alpines in the United States are descended from this group.

Size: Medium to large. 30 inches (76.2 cm) tall at the shoulder

Best use: Milk and cheese. This breed has well-shaped udders and teats, allowing for easy hand milking.

Breeding: Seasonal breeders. Alpines breed in the fall, giving birth in the spring. Alpine kids can have different coloring than their parents.

Personality: Big and proud animals. (Some may even say headstrong.) They are highly productive

Suitable for backyard goat farming? Yes. Alpines adapt easily to most living situations while maintaining good health and steady milk production.

LaMancha

Nigerian Dwarf

LaMancha

LaMancha goats are easily recognized for their apparent lack of ears. Their ear flaps are very tiny—measuring less than 1 inch (2.5 cm) for "gopher" type ears and less than 2 inches (5 cm) for "elf" type ears. LaMancha are compact animals and come in any color or combination of colors.

Origin: They are thought to be descended from the short-eared goats of La Mancha, Spain, which were brought to California by Spanish missionaries. The modern LaMancha is a cross of the Spanish and California brush goats, first introduced to the dairy goat world by Fay Frey in California in the 1900s. It is considered the only true All-American dairy goat.

Size: Medium. 27 to 29 inches (68.6 to 73.7 cm) tall at the shoulder

Best uses: Milk. Although the LaMancha's milk production is slightly less than the Swiss breeds, the milk is higher in butterfat and sought after by cheese makers.

Breeding: Seasonal breeders, late July to mid winter. Kids can be of any color or color combination.

Personality: Friendly and easy to work

Suitable for backyard farming? LaManchas make for a great small-holding goat. Be ready to answer the question "What happened to her ears?" over and over again.

Nigerian Dwarf

The Nigerian Dwarf goat is a miniature version of her larger Swiss breed cousins and can have any coloring.

Origin: They are from West Africa. Legend holds that that they traveled in cargo ships carrying lions for zoos and were used as food for the lions during the trip.

Size: Miniature. The average height, measured at the shoulder, is just 17 to 19 inches (43.2 to 48.3 cm) for does and 19 to 21 inches (48.3 to 53.3 cm) for bucks.

Best use: Milk and pet. Milk production averages two quarts (1.9 L) per day with the highest butterfat content of all breeds (more than six percent). The Nigerian Dwarf is a good choice for small cheese production. Be sure you can work with her small udder. Nigerian Dwarfs are naturally docile and perfect as a child's companion animal.

Breeding: Able to breed throughout the year. Many births are twins, and the kids can be any color.

Personality: Friendly and easy to handle. They are easily adaptable to most living situations.

Suitable for backyard goat farming? Yes. Nigerian Dwarfs' small size means housing and food needs are also smaller than average goats—three Nigerian Dwarfs have the same needs as one normal-size goat—and their milk production isn't overwhelming. Their delightful character is a bonus.

Nubian doeling

Nubian

Nubian goats have long pendulous ears and pronounced Roman noses. They are tall and meaty goats in all shades and color pattern, even spots.

Origin: They have origins in Africa, India, and the Middle East, crossbred with native milking goats in the United Kingdom. They are also known today as Anglo-Nubian goats.

Size: Large. Some family lines are particularly tall, 30 to 35 inches (76.2 to 88.9 cm) at the shoulder.

Best use: Milk, cheese, and meat. Nubians are dual-purpose goats, but most Nubians are kept for milk with meat being secondary. Nubian goats have well-formed udders that allow for easy hand milking. The Nubian's average milk production per year is lower than many dairy breeds, but the butterfat content is among the highest, an important consideration for cheese making.

Breeding: As with most southern hemisphere goats, Nubians are able to breed outside of the late July to mid-winter season of the Swiss breeds, thus extending your milking season (if desired).

Personality: Very friendly and giving. They have the reputation of being quite vocal, meaning they like to communicate (bleat) frequently.

Suitable for backyard goat farming? Possibly. Nubians do well in a small space. Keep in mind how vocal this breed is and how close your neighbors are.

Oberhasli buck

Oberhasli

Oberhasli (pronounced *oh-ber-hahs-lee*) are beautiful animals with distinctive breed markings. They have striking reddish coloring with black dorsal striping and leg and head trimming. Oberhasli—Obi, for short—originated in mountain terrain and are known for being strong for their average size.

Origin: They are originally from central Switzerland. Oberhasli were imported to the United States in the early 1900s.

Size: Medium. At 28 inches (71.1 cm) shoulder height, Oberhasli are slightly shorter than the other Swiss breeds.

Best use: Milk. Many say the milk of the Oberhasli is the sweetest tasting of all goat milks.

Breeding: Seasonal breeders, late July to mid winter

Personality: Strong herd mentality (easy to keep together). They tend to be herd bosses when mixed with other breeds.

Suitable for backyard goat farming? Yes. Their gregarious and giving nature and medium size fit well into limited space.

Saanen

Toggenburg

Saanen

Saanen (pronounced sah-nahn) goats are the largest of the dairy breeds and the most prolific milkers. Saanens are short-haired, though they may have longer hair along the ridge of the back and the top of the back legs and are pure white with a very feminine look. Goats that look like Saanens but have cream or gray coloring are referred to as Sables. The difference is in color only.

Origin: They are from the Saanen Valley in the south of Switzerland near Bern.

Size: Large. 30 inches (76.2 cm) tall or taller at the shoulder.

Best use: Milk. They have well-formed, easy-to milk udders.

Breeding: Seasonal breeders. Saanen breed in the fall and give birth in the spring. Saanen kids are especially eye-catching.

Personality: Friendly and curious, like all Swiss breeds.

Suitable for backyard goat farming? Yes. Although slightly larger than the other Swiss breeds, Saanens adapt well to confined spaces and thrive in large herds or small groups.

Toggenburg

Toggenburg are handsome, compact animals, with light fawn to dark-chocolate coloring and distinctive white markings down each side of the face, on the legs below the knees, on the lower back legs below the hock, and on both sides of the tail.

Origins: The oldest known breed of dairy goat, it is from the Toggenburg valley in Switzerland.

Size: Medium. The smallest of the Swiss breeds, Toggenburgs are 27 to 30 inches (68.6 to 76.2 cm) tall at the shoulder.

Best use: Milk. Toggenburgs have well-formed udders and are bountiful producers of milk lower in butterfat than the other popular dairy breeds.

Breeding: Seasonal breeders. Toggenburgs breed in autumn and give birth in spring.

Personality: Some say they are the least docile of the Swiss breeds, but are easily adaptable to all adequate housing situations.

Suitable for backyard goat farming? Yes. Their somewhat smaller size and productivity make them excellent backyard animals. They handle cold climates better than most other breeds.

Milk Production and Butterfat Content

This chart lists the major dairy breeds' annual average milk production and the milk's butterfat content. *Note: Your goats may give slightly (or significantly) more or less depending on breeding, care, and feeding.*

Breed	Milk Production	Butterfat Content
Alpine	286 gals. (1,083 L)	3.5%
LaMancha	257 gals, (973 L)	4.2%
Nigerian Dwarf	N/A	N/A
Nubian	219 gals. (829 L)	4.7%
Oberhasli	276 gals. (1,045 L)	3.6%
Saanen	313 gals. (1,185 L)	3.2%
Toggenburg	270 gals. (1,022 L)	3.1%

*These numbers are based on average milk production. Comparable testing results are not available for Nigerian Dwarf goats.

Other Dairy Breeds

Other, less common dairy breeds include the following:

- The Kinder, a cross between the Pygmy and Nubian breeds
- The Poitevine of Western France, whose milk is used in making many French cheeses
- The long-haired Golden Guernsey from the British Channel Islands, known for their delicious milk
- The Maltese of Malta, who produce high butterfat milk
- The Australian Melaan, a black version of the pure white Saanen

Meat Breeds

Goats in the meat breeds category have stronger builds and more meat on their bones than dairy breeds, which are more slender and angular looking. Two important traits to look for in a meat breed are reproduction and growth rate.

When raising meat goats, the product is meat. The more kids a doe produces over her lifetime, the more product you will have to sell or consume. Breeding year round gives the possibility of having new kids on the ground every eight or nine months compared to the once-a-year kiddings of most of the dairy breeds.

Meat breeds also have an accelerated growth rate when compared to dairy breeds. The high butterfat milk and genetic makeup of the meat group ensures fast-growing kids that put weight on quickly. The faster the kids get to market weight, the faster you can sell them and the more efficient your operation.

Goats raised for consumption are slaughtered as early as six weeks and up to one year after birth. Of course, all goats can be sold to a butcher for meat at the end of their "useful lives," the farmers' term for an animal's years producing milk or fiber or as breeding stock. Non-meat breeds are usually sold for a much lower price, which farmers call "salvage value."

Goat Keepers' Glossary

Hock: This is the joint in the goat's back leg which bends in the opposite direction of the knee on the front leg.

Boer

Kiko

Boer

Boers are a handsome, large, meat breed with distinctive markings: an all-white body and brown or red head. Boers have long, pendulous ears and a very stocky build.

Origin: They are originally from South Africa. The Boer goat was bred for fast meat growth and resistance to disease.

Size: Large. These are heavy goats, built like a tank, with females reaching weights of 200 to 220 pounds (90.7 to 99.8 kg) and males weighing in at 240 to 300 pounds (108.9 to 136.1 kg).

Best use: Meat. Fast growing and powerfully built, the Boer is an ideal meat goat. Boer milk, although high in butterfat, is used exclusively for growing kids as milk production quickly tapers off near the end of the weaning period (eight to twelve weeks).

Breeding: They are able to breed year long with the possibility of three pregnancies every two years. Boers mature quickly and are ready for breeding at five months.

Personality: Friendly, docile, and big. They are known for superior mothering skills.

Suitable for backyard goat farming? Possibly. Boers' friendly nature makes them easy keepers, but their formidable size and power may be a drawback in small spaces.

Kiko

Native to New Zealand, Kiko goats are large framed with sizeable, graceful horns. Although generally white, they do come in all colors. Kiko have pronounced weight gain with minimal care and are able to thrive on their own given plenty of acreage with large plant diversity. Kikos are parasite- and disease-resistant and thrive in a variety of climates.

Origin: Feral goats running in the wilds of New Zealand were bred with domestic dairy goats to create the Kiko breed.

Size: Medium to large. About 32 inches (81.3 cm) tall at the shoulder with a meaty carcass.

Best use: Meat. Kikos achieve weight gains at rapid pace, with does reaching up to 150 pounds (68.1 kg) and bucks from 250 to 300 pounds (113.4 to 136.1 kg).

Breeding: They are able to breed all year, which allows for more than one breeding per year (about every eight to nine months), producing rapidly growing, vigorous kids.

Personality: Kiko bucks are proud with plenty of personality. Kiko does are graceful and feminine.

Suitable for backyard goat farming? Not recommended. Given the proud nature of the buck, small confinement would be challenging. Kiko goats are best suited to larger spaces with room for exercise and plenty of opportunity for making use of their foraging ability.

Pygmy

Pygmy

Pygmies are short and compact animals that come in all shades of color, with many breeders favoring shades of gray or brown. Although short in size they move gracefully and are fun to watch.

Origin: Equatorial Africa

Size: Small. Does measure 16 to 20 inches (40.6 to 50.8 cm) to the shoulder; bucks, about 2 inches (5 cm) taller.

Best use: Meat, milk in small quantities, and pet. Milk is high in butterfat and cheese making is possible, albeit in small quantities.

Breeding: Year-round breeding. Triplets are not uncommon.

Personality: Pygmies, with their compact size and fun personality, not only give us entertainment but economical weed control.

Suitable for backyard goat farming? Yes. Pygmy goats are delightful additions to the backyard farm.

Myotonic
Right: Myotonic "fainting"

Myotonic

Myotonic goats are better known as Tennessee Fainting Goats. They come in any color or color combination and have crimped ears. Blue eyes are not uncommon, but their most notable feature is their "fainting." When startled, the muscles in their legs stiffen, causing them to fall over, but not actually faint. This is a genetic condition called myotonia.

Origin: A farm worker from Nova Scotia reportedly ended up in Marshall County, Tennessee, with four of these stiff-legged goats in the 1880s. They were bred by a local fancier for their prolific breeding and fast weight gain—and for their inability to escape enclosures.

Size: Small in height, small to large in weight. Averaging 17 to 25 inches (43.2 to 63.5 cm) tall, Myotonics range in weight from 60 to 160 pounds (27.2 to 72.6 kg).

Best use: Meat and pet. With their frequent "fainting," Myotonics are a popular breed for hobbyists.

Breeding: An extended breeding season allows for frequent births with a high percentage of twins.

Personality: Easy to manage and friendly

Suitable for backyard goat farming? Yes. They are easily fenced and fun, but also easily—and dramatically—startled.

Keeping Goats for Meat

If you breed your goats, you will most likely have kids that will be sold or consumed as meat. It is reported that goat is the most consumed meat worldwide. Low in cholesterol and fat, goat meat is historically a favorite in many regions of the world and is sought after for religious ceremonies. Backyard goat farmers can help raise awareness of this healthy (and "local") food choice.

Selling unwanted kids for meat can also be an important boost to your farm economics. For example, selling young twins for meat can help pay for the hay requirements of your doe for the year. Ethnic markets are a good place not only to sell your kids but also to find a butcher, should you choose not to do it yourself.

Goat kids mature quickly and are large enough to slaughter as early as six weeks. At nine weeks, a kid from a standard size goat will dress out to 25 pounds (11.3 kg) or more. (Hanging weight, or dress-out weight, refers to the weight of the consumable carcass, meat, organ meats, and bones. Hanging weight is estimated at about 60 percent of an animal's live weight.)

Before you decide to butcher at home, make sure it is legal and appropriate to slaughter in your backyard. If you live in a densely populated area or you will be slaughtering outdoors where passersby may witness the act, your home may not be an appropriate location. You must operate with respect to your neighbors and to the animal at all times.

Fact: To have meat, there must be a kill. It is your responsibility as a goat farmer to make it as quick and humane as possible. This book does not cover butchering, but you can learn about several methods of slaughtering from other sources. Know that blood, entrails, and skin must be dealt with appropriately to avoid flies and scavengers. Butchering should be done during the cool hours of the day. Meat should be handled cleanly and quickly to get it cooling to prevent spoilage.

Fiber Breeds

Fiber breeds are those goats whose main product is hair or fiber, specifically cashmere or mohair. Cashmere is a type of downy hair used as insulation in some sixty-eight breeds of goats around the world. Mohair, derived from the Arabic word for goat hair, comes exclusively from Angora goats.

Cashmere is combed out from the goat; mohair is sheared. Mohair is attractive to spinners and weavers because of its long, high-quality fibers, which readily absorb dyes and retain color well.

Keeping Goats for Fiber

If you are interested in having more than a pet goat but are not looking forward to daily milkings or raising kids, fiber goats may be a good choice for you. Fiber goats require less daily commitment than dairy goats, and if you keep castrated males, you will not have to deal with the yearly breeding, birthing, and possible butchering.

Fiber goats grow long, luxurious coats that need to be shorn only twice yearly. If you keep only a couple of them, your annual shearing work will not take very long.

Angora goat

Angora

Medium size with long, curly hair fibers, droopy ears, and long horns, Angora goats are the producers of the high-priced fiber mohair. Though white was the preferred color for many years, a concentrated effort is now being made to breed grays, blacks, and browns.

Origin: Ankara Province, Turkey

Size: Medium. 22 to 25 inches (55.9 to 63.5 cm) at the shoulder

Best use: Fiber. Under ideal conditions, fiber grows up to an inch (2.5 cm) per month, allowing two shearings per year of 4 to 6 inches (10 to 15 cm) each. The best mohair is obtained from kids and first shearings. As the Angora ages, its fiber becomes coarser. Wethers are excellent producers of fiber, as they do not have the stress of breeding or kidding.

Breeding: Seasonal breeders. Late July to midwinter with single kids not uncommon

Personality: Friendly and easy to manage, although not considered as naturally hardy as other breeds

Suitable for backyard goat farming? Yes. They are ideal for backyard farmers and weavers but do not do well in hot, humid conditions. They are easy to maintain with very little work (apart from kidding and twice-a-year shearings).

How to Hand-Shear Goats

To hand shear with manual shears, depending on how easy to handle your goat is, you may be able to simply tie her up. The common position for shearing is to "sit" the animal on her backside, keeping her sitting up between your legs. If done correctly, it will cause minimal protest. Snip the fiber carefully and quickly.

When shearing, always keep the sharp points away from the animal and yourself.

The first time shearing with hand shears will take some time until you get the hang of the technique. For the comfort of the animal (and your back), work confidently, quickly, and rhythmically until the animal is shorn. Start at the head and work your way down her body, keeping her skin taut with one hand while shearing away from you with the other to avoid snips (to your goat and yourself!). Electric shears, although costly, cut your shearing time to mere minutes.

Once shorn, the fibers must be sorted into usable and nonusable fibers (overly soiled or snagged fibers should be discarded or composted). Usable fibers are washed, dried, and spun. Many references are available for learning how to spin raw wool. If weaving, knitting, or other fiber arts interest you, sweaters and scarves made from mohair will make gifts or saleable items.

CHAPTER CHECKLIST

Have you...

- ☐ Considered your goal in goat keeping—milk, meat, fiber, or companionship?
- ☐ Researched the breeds that will fit your goal? (Visit the goats on a farm if you can.)
- ☐ Decided on a goal-appropriate breed, factoring in the breed's personality, physical traits, and needs?

From the Farm

A Lesson in Milking

Twenty years ago, I bought my first sheep. I had just arrived in Tuscany, and I saw little flocks of sheep scattered throughout the hills, tended by shepherds. That was exactly the life I wanted for myself, but although I had a lifelong passion for animals, I knew nothing about sheep.

One cold day in February along with my friend Lea, the manager of a high-end Manhattan restaurant, I went sheep shopping. Faced with dozens of large, woolly animals, I asked Lea in my best farmer voice, "What do you think?"

She lifted her sunglasses and blurted out, "I don't know!" I didn't know either. So I bought forty of them.

The sheep were trucked to my farm on the mountain, and after a few days of getting to know each other, I decided it was time for milking. Out of forty sheep I don't believe I got any more than a few liters of milk. I did get plenty of sheep poop in the bucket, sheep feet in the bucket, and handfuls of sheep wool in the bucket as the sheep pulled away from me, not wanting any part of my Tuscan dreams.

I had bought Bergamasca crossbred sheep—the largest breed of meat sheep in Italy. Before arriving in my makeshift barn, these sheep had never been milked. Still, I milked them every day and made cheese with that little bit of milk every night.

Passing Italian farmers stared at me while I was milking. They knew something I didn't: If you want milk, milk a dairy sheep, not a meat sheep. Eventually I sold my Bergamasca meat sheep for a purebred flock of Sardinian milk sheep.

Milking my meat sheep gave me a bucket of milk. Milking the same number of dairy sheep gave me four buckets of milk.

A pair of handsome goats enjoying their backyard "perch"

CHAPTER 4

Goat Needs

Your work as a goat farmer starts long before you bring home your first goats. Preparation is key, whether you are on a spacious farm or in a cozy backyard.

Look at your space through the eyes of a goat. How can you best meet her needs for adequate space and safe, clean housing? How can you meet your needs for a strong fence and easy-to-clean structures? On a farm, the goats are probably in a comfortable corner of a lofty barn with double-wide doors leading to a rolling pasture—or at least that's how we like to imagine it. The life of the goat and the goat farmer in a backyard setting is different. You will encounter some issues not usually faced on a farm, including storage space concerns and the proximity to neighbors. In a backyard, you can't replicate the farm experience, but with some creative thinking and advance planning, you can create a good, healthy environment for your goats.

This small backyard goat enclosure is a seamless, pleasant addition to this dense neighborhood setting.

Evaluating Your Space

Do you have the space to keep goats? To raise happy, healthy goats, you will need room in your backyard for a goat pen and a goat house, as well as storage space for the goats' food and other goat-related supplies such as straw. (See chapter 5 for more details on feeding your goat.) And you must know what to do with all that soiled goat bedding that you will clean out at least two or three times a year.

Backyard Planning

How much room you need will depend on how many goats you will house. Each goat should have ample floor space for sleeping, generous space at the feeding trough, and access to an outdoor enclosure. If you are breeding your goats, keep in mind that your herd could double or triple each spring. (For more details, see "Backyard Housing" in this chapter, page 48.)

- The **placement of the goat enclosure** within your backyard is also an important decision. In a farm setting, the placement of the enclosure is determined more by convenience than science; it may be simply a fenced pasture attached to the barn. In a smaller setting, you'll have to think more carefully about where you are going to set up your goat area, giving consideration to your goats, your neighbors, and of course, yourself.

- A goat's pen should have areas of both **sun and shade**, which can be provided by an overhang attached to their housing or a large tree.

- It should have **protection from strong winds**. The goats' pen should be attached to the goats' house, where they will also seek shelter from the elements.

- The goats' pen should be **free of ornamental plantings and exotic grasses**. Other plants such as ferns, rhododendrons, azaleas, and mountain laurel are poisonous to goats. If there are young trees in the enclosure, they should be well protected (that is, fenced); otherwise, the goats will make a quick meal out of them.

  Plants may be poisonous or cause negative effects for several reasons, and the level of toxicity depends on several factors, including the stage of growth of the plant, which part of the plant the animal consumed, how much was ingested, and for some plants, at what stage of decay the plant was eaten.

The Danger of Toxic Plants

Cyanogenic plants such as milkweed, mountain laurel, pit stone fruits, and leaves interfere with the blood's ability to carry oxygen. Death is usually very rapid. Photodynamic poisoning is usually more of a concern for animals with areas of unpigmented skin. Rape and St. John's wort are photodynamic plants. When they are eaten in large amounts, sores develop on the skin when exposed to sunlight. Other plants, such as ferns, if consumed in large amounts may cause internal hemorrhaging.

Contact your local health department or agricultural extension agency for a list of poisonous plants in your area.

Common Poisonous Plants

Fern (*Filicopsida*)

Rhododendron (*Rhododendron catawbiense*)

Azalea (*Rhododendron obtusum*)

Mountain laurel (*Kalmia latifolia*)

Goats explore with their mouths, and their targets may include your fencing! Keep their safety (and yours) in mind when considering fencing options.

Backyard Fencing

Building and maintaining good fencing is very important. Goats are escape artists. You need strong fences, at least 4 feet (1.2 m) tall, to keep them in, and goats have been known to crawl under a fence with a gap of less than 7 inches (17.8 cm). Fences are also necessary to keep predators out.

There are several things to consider when choosing fencing, including the area to be fenced, the cost, and the ease of maintenance. Depending on where you live and how close your neighbors are, you may also want to consider the attractiveness of the fencing. Remember: Fencing needs to be built before you bring your goats home.

Tip

If your backyard is small and your neighbors are close, take extra precaution with the placement and design of your goat pen and house to protect your neighbors from unwanted sights, sounds, and smells. The secret to keeping happy neighbors is to invite them to participate in all the good that goats offer (playful antics, cute new babies, fresh milk, and a wheelbarrow full of wonderful compost) without making them participate in the not-so-good (smells, noises, and escaped goats who eat hedges, fruit trees, and flowers).

Protection from Predators

Your housing and fencing must protect your goats against predators—and the biggest threat to your backyard goats may be the neighborhood dogs. Dogs, by instinct, are chasers, and even your well-behaved pet may turn into a hunter when confronting a skittish goat. Your goats need to feel safe from these and other dangers within the enclosure you have provided for them to be able to produce their best for you. If your goats are subjected to a dog of any size running back and forth along the outside of their enclosure all day, the goats may spend their time cowering in the corner wondering if that barking mutt is going to get through the fence!

Goats, especially those in small spaces, consider anything that threatens them as a predator—and that can include children. For that reason, goats and children should not be left alone together until both can be trusted.

Other, more traditional predators to consider depending on where you live include coyotes (increasingly more common), bears (rare), wolves (limited areas), and large cats, such as mountain lions west of the Rocky Mountains. If you live in areas where these predators are present, consult an experienced goat or sheep keeper as to how to best protect your animals. Other common animals like raccoons and skunks are not a threat to goats, though they can be a threat to the goats' feed. Be sure that feed containers are goat- and rodent-proof.

Be aware of potential predators, such as this coyote (*Canis latrans*) in your area.

Types of Fencing

Woven Fence

Woven wire netting, with square openings less than 4 inches by 4 inches (10.2 cm by 10.2 cm), is nailed to sturdy wooden posts placed 6 feet (1.8 m) apart.

Pros: Long-lasting, if well built

Cons: A small weave will lessen the chance that a full-grown goat will get her head stuck, but kids need extra supervision, especially if they have horns. The fence can sag if not strung tightly.

Cost: Midrange

Suitable for backyard goat farming? Yes, highly recommended.

Board Fence

Woven Fence

Board Fence

Wooden boards are nailed horizontally between sturdy wooden posts.

Pros: Beautiful to look at

Cons: Unless you use many boards with only small gaps, it is not likely to keep in a goat that wants to be on the other side. (And most goats want to be on the other side.)

Cost: Expensive

Suitable for backyard goat farming? No. The likelihood of escape is too high.

Electric Fence

Strands of electrified wire which may be intertwined with a plastic line strung 10 to 12 inches (25.4 to 30.5 cm) apart horizontally between posts, give a shock to any goat (or other animal or human) that touches it.

Pros: It's the cheapest option for larger or difficult-to-fence areas; effective once the goats learn to respect it.

Cons: The electric wires must be monitored to keep them free of grass, fallen branches, broken fence posts, or anything that might lessen the charge.

Cost: Inexpensive

Suitable for backyard goat farming? Not recommended for small spaces. It is ineffective at keeping predators out. It is shockingly frightening (literally) to small children who may come over to visit the goats in your absence. Electric fencing can be suitable for larger spaces where concern for predators is low.

Barbed Wire

Strong metal wires, usually two strands wrapped around one another for added strength, have sharp barbs placed at short distances to deter animals from attempting to get past it.

Pros: None.

Cons: Barbed wire is not effective at keeping predators out and is very dangerous for your goat, who can suffer permanent damage from torn eyelids, ears, and udders should she try to get through it. It is also dangerous for children

Cost: Inexpensive

Suitable for backyard goat farming? No.

Barbed Wire

Steel Stock Panels

These are large solid metal panels in 10- to 30-foot (3 to 9 m) lengths. Panels are designed to hook together, and four panels make a small square pen.

Pros: Easily set up and easily moved

Cons: Unattractive and costly for fencing small areas.

Cost: Expensive

Suitable for backyard goat farming? It's suitable for keeping goats in, but you should go have a look before deciding to line your backyard with them!

Steel Stock Panels

Hybrid

This is a combination of the woven wire and board fencing methods. Boards are placed on the inside (goat side) of a tightly stretched woven fence.

Pros: The wire prevents goats from escaping under the boards, and the boards prevent the wire from sagging.

Cons: None

Cost: Medium range. But if constructed correctly, it will last for years and is very predator safe.

Suitable for backyard goat farming? Yes. This is the most highly recommended and functional fencing method.

Traditional barns were built to be sturdy and warm and with easy access to pastures.

Remember, just because it might work for your goats, you should ask, is it workable for you?

Backyard Housing

Goats are not too particular about their housing. Plenty of happy goats live in what could, at best, be described as rudimentary housing, and plenty of happy goats live in dairy barns that could be described as luxurious. As long as they are safe and have adequate room and protection from the elements, your goats will adapt well to any housing you can supply. Remember, just because it might work for your goats, you should ask, is it workable for you? In addition to the goats' safety and comfort, you will want to construct housing that makes it easy for you to do your daily chores, with high ceilings and wide doors or aisleways.

Two's company. The first thing to consider when constructing housing is the maximum number of goats you will have at one time. All the goats need enough space to lounge about without being trampled on. They need individual space at the hay feeders to ensure they get their share and access to clean water at all times.

- A 4-foot by 9-foot (1.2 by 2.7 m) stall is ample space for a goat and her kids, providing they have access to a yard. Two does need about 70 square feet (6.5 sq m) of housing space.
- Your goats' housing should be connected to securely fenced outdoor space as large or small as you can accommodate.
- Use common sense when deciding your goats' spacing needs. If their housing is open and airy and connected to a big yard, you can get by with a smaller footprint in their housing. If the outdoor area is very limited, you may need to give them more space indoors, as that is where they will be spending the majority of their time.

Safety and Security

If predators of any kind are a threat, you need a way to close your goats safely inside their housing when you are not around to supervise. If you are considering windows in their housing, make sure they are above the head of the tallest goat when she is standing on two feet and leaning against the wall (about 5 ½ feet [1.7 m]) or covered with bars or tight screens so she can't accidentally poke her head through, break the glass, and cause injury.

Visiting Hours

You will be going to the goat barn a minimum of twice a day. The easier you can make it on yourself and the more comfortable it is to work, the more enjoyable your goat experiences will be. You will need to clean out your goat barn several times a year at least. Having enough headroom to stand up straight will make this job much more tolerable and less backbreaking. Wide doors or double doors to get your small garden tractor or wheelbarrow through are always a luxury come cleaning time.

Three examples of small-space goat housing: All are safe and comfortable and blend in nicely with their surroundings.

Providing a deep layer of straw in your loafing area keeps your goats comfortable and dry for months at a time.

Bedding

Your goats will be warmer, more comfortable, and healthier if they have a nice layer of bedding to lie on in their housing. Bedding is an organic, absorbent material that builds up over time as fresh material is piled on top of soiled bedding. The barn must be cleaned out at least several times per year. Goats are not particular about where they leave their droppings, and we don't want them to be sleeping in soiled bedding, so keep an eye on manure buildup and be sure to cover with a fresh layer as needed.

Straw

Straw is the cheapest choice for bedding. Straw absorbs urine and droppings easily. Start with a big fluffy layer of clean straw, which will allow urine and droppings to find their way to the floor, leaving the top layer clean and comfortable. As the bedding starts to get dirty, just put on a fresh layer of straw. For your goats' health, do not make them sleep in moist, soiled bedding. By continually adding fresh straw at the first signs of soiling, you can build the bedding for months on end without cleaning, and the barn stays fresh looking and smelling. This is called the deep-pack method. The downside of this is that when you do clean it out, it is deep and it is packed! You will need to do a full cleanout at least twice a year, possibly more, depending on the number of goats you are keeping, the size of their loafing area, and access to outside pasture. Turning up a stall full of deep-packed bedding will obviously be aromatic, which your neighbors may not appreciate.

Straw bales are about the same size as hay bales. You will need to have about 25 square feet (2.3 sq m) of storage space, 8 feet (2.4 m) high, to stack twelve straw bales, which will last about three months, depending on the size of the housing and how much time your goats spend outside.

Wood Shavings and Sawdust

A bed of wood shavings and sawdust is similar to a bed of straw. You add a good fluffy layer of sawdust frequently to keep your goats dry. It looks great when fresh, but it must be cleaned more often than straw. Wood shavings, after being urinated on regularly, will mat together and not allow the droppings to fall through. Be careful when using a sawdust base to your bedding, as wet sawdust is found to have large numbers of bacteria, which can cause a type of mastitis. (See chapter 11 for more information on mastitis.) For this reason, when using sawdust, more frequent cleanings, which are easier on the back, are recommended.

Goats seem equally happy when they have bedding of fresh straw or new wood shavings.

Cement Floors

Most barns or animal housing will have floors made of dirt, packed clay, wood, or cement. Normally a layer of bedding as described above is put on the floor, but some goat owners prefer to leave the flooring bare so it can be swept out daily. Most goats, if given the choice, will sleep elevated off the ground. If you have slightly raised platforms around the walls of the housing, the goats will use the platforms for resting, making daily sweeping easy. Daily sweeping also cuts down on flies and saves you from big barn-cleaning days. As always, use good judgment when selecting bedding. In a mild or warm climate, sleeping platforms can be a good choice. However, in colder climates, a sleeping platform will not provide the warmth that the insulating deep-pack method will.

Food and Water

Goats need ample space at the feeding mangers. One of your goats will be the boss and may push the other goats away if she feels crowded while eating. With enough room at the feeder, your more submissive goats will be able to get their daily nutritional needs.

A hay feeder that is 3 feet (1 m) long will do the job for two goats and their eventual kids. As always, observation is the key, and if you notice that one of your goats isn't eating, reposition your feeders so the goats are separated or experiment with a bigger feeder.

One water source is sufficient for two goats. This can be as simple as a bucket in a corner or an automatic waterer attached to the wall. If you live in a cold climate, an automatic waterer with a heating element may be the best choice to prevent the water from freezing. Automatic waterers should be installed so that the cords are not exposed for the goats to chew on.

Goat Keeper's Glossary

Straw. This is the stem of grains such as wheat, oat, and barley. As the grains dry out and are ready to pick, a combine passes through the field, separating the grains from the tall, dry yellow stem. This is called straw (it looks like a straw—long and hollow) and it is baled like hay, making it easily stacked and stored.

Cleaning and Maintenance

Use common sense: when your goats' home looks dirty, it is dirty. For the well-being of your goats and your neighbors, regular cleaning is important.

Manure

The reality of raising goats is that you will also have to deal with manure management. You will have to clean out your goat housing completely at least twice a year. If you use a deep-pack system, the bedding will slowly accumulate upward while your doors and fences stay at the same level—it will seem like your goat has grown taller! If your bedding is mushy or so deep that your goats are looking over the fences, it's time to clean it. Even a small box stall measuring 10 feet by 10 feet (3 by 3 m) with deep-packed straw bedding 15 inches (38 cm) deep means nearly 5 cubic yards (3.8 cubic m) of manure that you will have to put somewhere. That creates a very large compost pile, which will give off odor for some time.

Be respectful of your neighbors and let them know your barn-cleaning days so they know that the smell of country goodness is temporary. It might also be a good idea to offer any neighbors who may be affected a few wheelbarrows full of your garden-perfect goat manure compost in exchange for their patience.

Assisting in Manure Management

Barn cleaning is a fact of life every goat farmer must endure. Much like going to the dentist, the anticipation is worse than the event. Many techniques make the labor-intensive task go smoothly.

- In large goat barns, you can employ a tractor with a forklift fitted on the front end for removing manure.
- Depending on the size of your goat pen, you may use a small riding lawnmower with a wagon attachment. The wagon is filled with pitchforks full of soiled bedding.
- You may use a wheelbarrow and walk it to a designated compost pile.
- Lining the back of a pickup truck with plastic and filling it up to haul it away is another option.

If you are friendly with other backyard gardeners and farmers, you may be able to enlist their help on cleaning day. Gardeners are interested in manure for compost, and by letting them take home what they clean out, you may not have to lift a pitchfork!

Pest Control

Keeping any farm animal on your property increases the probability of attracting flies and mice. A clean barn and some forethought can reduce this problem.

Controlling Flies

There are several ways to control flies, and the battle should start early in the spring, as soon as you notice the first fly Once warm weather arrives, flies are attracted to the heat of the barn, the smells, and the sweat from the animals.

Just add chickens. The most natural way to fight flies in the compost pile is to keep chickens on your property. If you have composted your bedding material, chickens will happily spend most of the day scratching through your compost pile eating worms, larvae, anything moving, which will cut down on flies, as their eggs are being constantly disturbed and eaten by the scratching chickens. An added benefit is that the compost will break down faster as the chickens move it about. They turn it for you, and they give you eggs in return.

Electric solution. If adding chickens to your backyard farm is not a possibility, insect zappers can be effective. These are electrical boxes with fluorescent lights that attract flying insects. The insects land on the electrically charged grill, which "zaps" them. Don't put the insect zapper in the goat pen, as the goats are a bit wary of the zapping—and they love to chew electrical wires. When placing your insect zapper, you must also keep it far away from hay or bedding, which could ignite in case of a spark.

Organic choices: Organic pest controls and home remedies may not smell the best, but they are effective. Garden supply, hardware, or farm and feed stores will carry organic fly attractant. You empty the all-natural contents into the plastic bottle and mix with water. The result is very smelly, which will attract flies by the thousands. The smell, though awful up close, is barely detectable to humans just a few yards away. A special lid on the bottle allows flies in and won't let them out. Hang in areas where you have fly problems. You can also make a similar fly trap with a soda bottle and something only a fly would find irresistible, such as a small piece of rotting meat. Cut the top off the soda bottle, put the piece of meat in the bottom, and invert the top half into the bottle so the flies crawl down toward the smell and are unable to crawl out.

Finally, fly strips—sheets of hard paper that are covered with a sticky coating and scented with male fly pheromones that attract and trap egg-laying females—are very effective when hung wherever flies are a problem. A word of caution: make sure fly strips are out of reach of goats, other household animals, and, of course people.

Goats are more than happy to share their space with your other backyard animals.

Your barn cat will help keep mice in check ensuring that your goat's food supply will not become contaminated by vermin.

Controlling Mice

Addressing mouse concerns is relatively easy. Avoid attracting mice in the first place by ensuring your goats' pellet food is well secured and not accessible. Large metal cans with tight-fitting lids work best. Be sure to clean up spilled goat pellets often and do not feed so much or in such a way that feed falls on the floor.

For controlling mice, keeping cats, especially female cats, in and around your property is the first choice.

A general word of caution when using other types of mouse traps: Always keep the safety of children, pets, and goats in mind. Never place traps where they may catch an unwanted victim, including inside the goats' pen.

Other options include standard mouse traps, glue traps, and poison. Regular mouse traps baited with peanut butter (not cheese, as dogs or cats may be tempted to eat it) need to be checked often to see if a mouse has been captured.

To use glue traps, a dab of peanut butter will attract the mouse, which is then stuck to the cardboard with sticky glue. This approach is effective, but you must kill the immobilized mouse with a quick thunk on the head before disposing, so it is not a very humane option. Humane traps, which capture the mouse in a wire cage, are available, but you must consider what you will do with the trapped mouse. Will you kill it, or will you drive it into the woods and release it?

Mouse poisons are sold in easy-to-handle packages. Poison is a torturous death for the mouse but an effective method of pest control.

From the Farm

Goats without Borders

When I started farming, fences were nonexistent and animal housing was whatever I could find. At first, my flocks of sheep and goats happily roamed the hillsides and woods without restriction. My farm stood more than a mile (1.6 km) up the Pico Della Castagna mountain from the public road that leads to the village of Caprese Michelangelo, with no neighbors living in any of the other long-abandoned farmhouses dotting the hillside. The sheep and goats would move constantly, sometimes traveling one or two miles (1.6 to 3.2 km) as they wandered, eating their fill and resting in the shade to chew cud.

From the top of the mountain, I could see the moving white flock. If they were close enough, I could hear their bells and know where they were and where they might be headed. As the sun was going down, I would set off towards where I had last seen my flock, listening for their bells. If I couldn't hear them, I would start looking for sheep and goat droppings. If they looked fresh, I would touch them to feel if they were warm, which meant I was getting closer.

I never lost my herd. But I would pick up my pace, my calling a little more frantically as the sun slipped away. Even if it was dark, I knew I would find them safe shelter. I was more concerned about finding my own way home.

If it was ever too dark to lead the sheep and goats home to Priello, I would lead them instead to the nearest abandoned house properties with names like Bencina, Capanno del' Lupo, Radicatta—and set up a makeshift gate to keep the herd safely inside and the wolves outside. I would go home for the evening and return in the morning to happy sheep and goats, ready for another day of roaming.

Today all of those abandoned houses have been renovated into luxurious vacation rentals or expensive homes, and when I pass by or go over for dinner, I always think about when my goats use to sleep in these Tuscan villas.

CHAPTER CHECKLIST

- ☐ Have you decided how many goats you will eventually house and determined that you have the necessary space?
- ☐ Have you chosen an area to fence that has protection from excessive heat, wind, and rain?
- ☐ Have you built a fence that will keep goats in and predators out?
- ☐ Have you checked with your local agriculture department for a list of poisonous plants for your area and made sure that those plants are out of goats' reach?
- ☐ Have you let your neighbors know of your plans to keep goats and how you are prepared to keep noise and smells to a minimum?
- ☐ Have you thought of a layout that is comfortable and safe for your goat and easily managed by you?
- ☐ Have you planned for manure disposal and pest control?

CHAPTER 5

Goat Feeds

There are two very important things to know about goats: They are browsers and they are ruminants.

Browsers, like goats, like to move about as they pick and select their favorite foods. Grazers, like cows, sheep, and horses, naturally lower their head and prefer to eat grass. So if you had hopes of your goats replacing your lawn mower, think again!

Their preferred eating area is from about 12 inches (30.5 cm) above the ground to as far as they can reach. Goats will feast on tall weeds, young trees, tree leaves, flowers, climbing vines, and bramble bushes. Keep that in mind when you take your goats out for a walk. Don't get too close to your neighbor's rose bushes!

Ruminants are animals, like goats, sheep, and cows, who ruminate and chew cud. Because goats have a complex four-chambered stomach, we need to take special care as to what and how much we feed them.

In a backyard setting, goats will most likely not be able to browse, so we as owners must provide them with their dietary needs, and we have to understand how the rumen works to keep our goats healthy. Therefore, providing your goat with a balanced diet is especially important.

Goats can and will stand on their hind legs to reach their favorite foods. Keep this in mind when building fencing and protecting trees and flowers.

The Rumination Process

Goats are capable of quickly ingesting large amounts of food, which they regurgitate to re-chew at a later time. After eating impossibly large amounts of hay or fresh greens, your goats will go through resting periods, either standing or lying, and happily stare into the distance and bring up their cud to re-chew for further digestion.

Some say that when you feed your goat, you are actually feeding her rumen, which, in turn, feeds her.

- The **rumen** is the first stomach, where freshly eaten, partially broken down food is stored. The rumen is the largest of the four stomachs with a holding capacity of between 3 and 6 gallons (11.4 and 22.7 L) depending on the size and breed of your goat.
- Food is regurgitated from the rumen and re-chewed before passing into the second stomach, the **reticulum**, where further material breakdown and digestion occurs.
- From the reticulum, food passes into the third stomach, the omasum. The **omasum**, which acts as a food grinder, further breaks down the food and allows more nutrients to be absorbed into the goat's body.
- The final stomach, the **abomasum**, acts much like the human stomach. Acids break down food not absorbed prior to arriving in the fourth stomach.

The digestion process in kids is different. When the kids are born, the rumen is much smaller than in adulthood. As they drink milk from a bottle or their mother, the milk is swallowed directly into the fourth stomach. As the kids slowly start to nibble hay and grains, the rumen will grow and the abomasum will shrink. For a doe, a large rumen may be an indicator of good milk production.

Complications of the Rumen

Of the goat's four stomachs, the rumen is the largest and most delicate. The rumen is filled with microbes that break down foodstuffs. It is essential to keep the microbes at a balanced level of activity. In the wild or in large areas, a goat will self-regulate to keep her rumen working properly. When goats are kept in a confined space, they have to rely on you for food. They will eat what you give them, and if you do not provide a healthy diet in correct proportions, the microbes will be overactive; the rumen can quickly go out of balance, causing bloat.

Bloat

Bloat is a fast buildup of gas in the rumen that the goat is unable to expel. Think of the rumen as a big fermentation vat. As the goat eats, partially chewed food is held in the rumen until it is rechewed and swallowed into the second stomach. As the food ferments in the rumen, it naturally produces gas, which under normal circumstances, a goat will expel through belching. Bloat happens when your goat overeats or is introduced to a new food in large quantities, causing an imbalance in the microbes in the rumen. The microbes begin multiplying quickly, overexpanding the rumen and impeding the goat's ability to expel gas through normal belching. A goat experiencing bloat may moan, hang her head, kick at her stomach, and foam at the mouth. Because the rumen is on the goat's left side, a bloated rumen will make her left side very tight, and it will appear rounded. Bloat is a very serious condition and without treatment can cause death.

Digestive System of the Goat

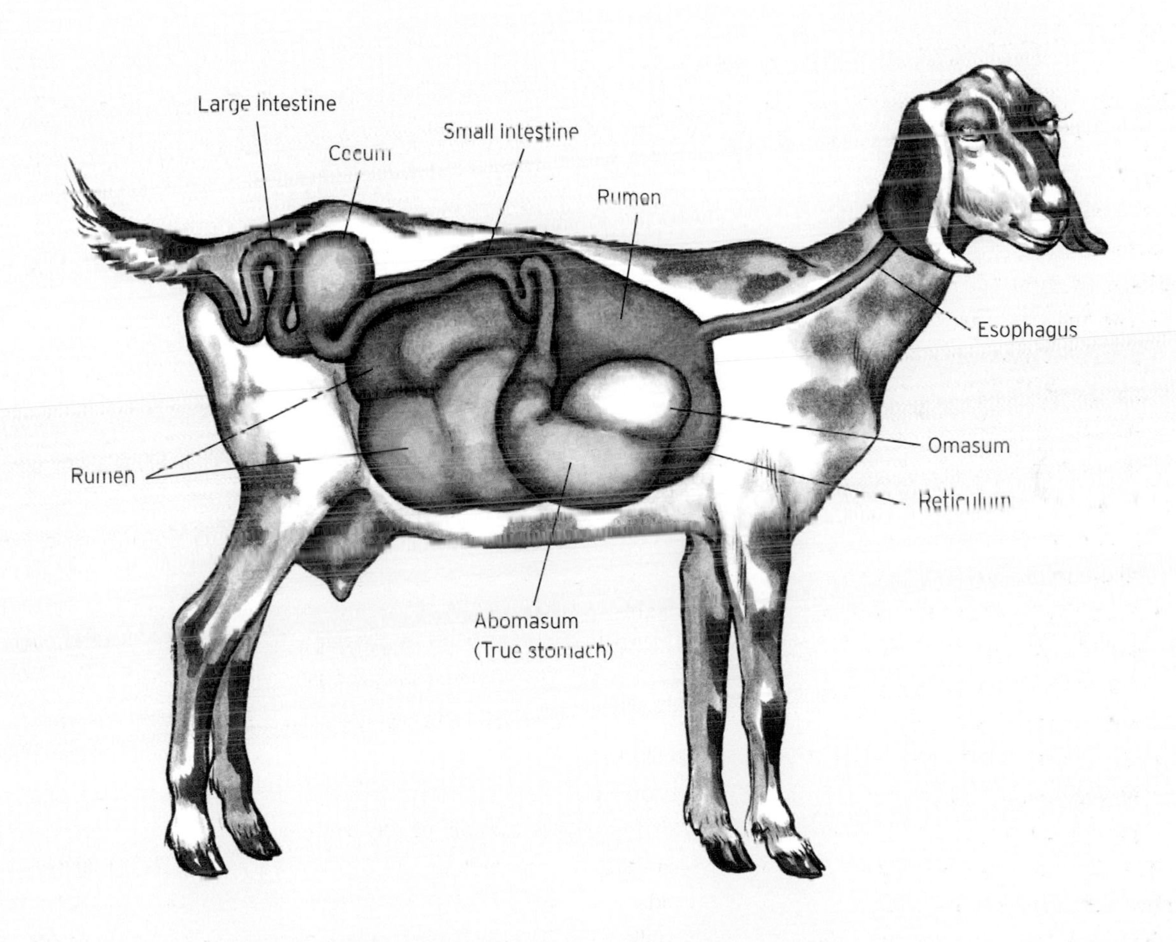

Hay

The main staple of your goats' diet will be hay. Hay is any grass or young grain (before the seed heads form) that has been cut and thoroughly dried. Adult goats will eat between 3 and 6 pounds (1.4 and 2.7 kg) of hay per day—all year long.

There are three types of hay: grass, grain, and alfalfa.

Grass hay is field and pasture grass cut at the flowering stage. Different regions will grow various types of grasses. Ask your haymaker or fellow goat farmers what types they have and why.

Grain hay is the grass stalks of barley, wheat, and oats in their fast-growing stage, cut before seed heads have formed.

Alfalfa is a fine-stemmed, branchy, many-leafed grass with a sweet smell. It is typically the most expensive of the hay varieties because of its nutritional qualities and high protein content. A diet of high-protein alfalfa, if combined with high-protein pellets, may cause problems such as diarrhea and excess energy if the diet is out of balance.

Hay season begins in late spring, when fields are mowed for the first time. Hay is categorized by cutting—first cut, second cut, third cut, and so on. The fast-growing second and third cuttings are generally considered to be the most nutritious hay, with first cutting yielding the most hay and including the greatest variety of grasses.

Estimating Hay Needs

Hay is sold by the bale. The typical bale is 45 to 50 pounds (20.4 to 22.7 kg) of hay. Some suppliers offer easier-to-handle 25- to 30-pound (11.3 to 13.6 kg) bales. Calculate your hay needs accordingly.

An average-size milk goat will eat one average bale of hay per week, so two goats will eat approximately 104 bales of hay per year, with consumption being noticeably higher in the winter and after giving birth and less in the summer. When estimating your hay needs, always err on the high side so you don't find yourself without hay with hay-cutting season still three months away!

The cost of a bale of hay varies widely from region to region and year to year. As they say, make hay when the sun shines—so if the sun isn't shining, your cost will go up. The farther hay has to be shipped, the more expensive it is going to be. Logically, the best, cheapest time to buy hay is in the summer when it is being cut. But when searching for your hay source—most often a local farm, feed supply, or garden center—be sure to ask if the hay will be available throughout the year. Don't wait until the snow falls to find out that the garden center charges extremely high prices or worse, doesn't carry hay at all in the winter.

Transporting and Storing Hay

The typical 45- to 50-pound (20.4 to 22.7 kg) hay bale is 18 inches by 2 feet by 4 feet (45.7 x 61 x 122 cm), an important consideration for transporting and storing hay. The trunk of an average car will hold two bales of hay. (Line the trunk with an old sheet to protect it from the tiny pieces of dried grass that will burrow into the carpet; they're almost impossible to get out!) An average pickup truck will carry twelve to sixteen bales, tied down tightly. Some hay sources will deliver the bales.

Cut waste by properly storing your hay and straw bales away from animals and the elements.

Storing hay can be a challenge in a small space. A year's worth of hay for two goats requires about 12 feet by 16 feet by 8 feet (3.7 x 4.9 x 2.4 m) of storage space. The space needs to be protected from both sun and water. You should not stack hay in a garage where cars will be running their engines. Exhaust fumes will soil the hay, making it unappetizing and unhealthy for the goats.

If you don't have room for storing a year's worth of hay at home, another option is to find a supplier who will store it for you.

Choosing Hay

Just like picking out your own fresh food, you want hay that looks and smells appetizing for your goats. The outside of a hay bale will fade with exposure to the sun, but the inside should have a nice green color and a grassy smell. Although the exterior of the bale may be damp due to rain or morning dew, which can be remedied with proper drying, the interior should be dry to the touch. Hay that is damp on the inside of a bale is hay that was not dried in the field properly and was baled before it was ready.

Visitors of all ages will enjoy interacting with your goats.

Refuse any hay that feels moist or hot on the inside. It may develop mold that is unappetizing to and dangerous for your goats. Moldy hay will be dusty and may have visible white growth or dark hay clumps. Damp hay is also a fire hazard. Moisture trapped within a hay bale will heat up like a compost pile, and spontaneous combustion can occur. If you suspect that a bale of hay is not dried, open it up and put your hand inside; it will actually feel hot.

If you mistakenly buy damp hay, open the bale, fluff it up, and spread it out in the sun. Let it dry out completely, and if it doesn't smell moldy, it most likely will be okay for your goats. This must be done immediately, as mold spores will grow quickly.

Feeding Your Goats Hay

Goats should have access to hay at all times. Goats will also waste hay if given large amounts of it as they pick and choose their favorite parts and disregard the rest. If your goat time is limited and it's easier for you to provide a large amount of hay for your goats in the morning so they can feed on it all day, you can expect more waste as a trade-off for less time involved in chores. If you can feed lesser amounts more often throughout the day, your goats will probably waste less, but you are dedicating more trips to the goat shed.

Design your hay feeders in a way that it's not easy for your goat to pull out large mouthfuls at a time, inevitably wasting hay as it falls to the floor. You can do this either by using what are referred to as keyhole feeders, so the goat has to put her head through a "key" hole to reach the hay, or by building your hay feeder with the slats close together so the goat can pull only small mouthfuls out at a time.

How to Make Your Own Hay

If you have a small yard or garden that will grow tall grasses without fertilizer or pesticides, you can make some of your own hay.

To make hay let the grass grow to a height of at least 1 foot (30.5 cm). The higher (within reason) the grass, the more hay you are able to make. You want to cut your hay before the seed heads have flowered and dried.

Day 1: On a sunny, dry day cut the grass with a scythe or weed trimmer, making nice, even cuts at ground level to produce long pieces. Let the cut grass dry in the sun for a few hours and then walk through with a rake or pitchfork and fluff up the grass trimmings. The goal is to dry the grass as fast as possible.

Day 2: The second day after cutting, after the morning dew has dried off, fluff up the cut grass again, separating any clumps. Let the grass, which will now start to resemble hay, rest in the sun until midafternoon. When you still have a few hours of full sun left, turn all the grass again. If the weather has been cooperating, by the end of the second day your hay should be very dried out and almost ready for storing.

Day 3: Flip all the trimmings again around midmorning, making sure any morning dew is completely dried off. By midafternoon—about sixty hours after cutting—you have hay that is ready for storage. To be sure hay is cured, grab a handful and clench it in your fist. It should feel dry and crinkled. Break the hay in half—there should be no bright-green sign of fresh grass. If you are unsure whether it is dried enough, let it sit another day, flipping twice a day as before. Remember that any hay that is not completely sun cured may go moldy, which is unsuitable for your goats. It's a lot of work to cut and flip and rake and put away only to find out it wasn't quite ready!

Gather the hay gently with a rake. Tie the hay in a bundle and store under a cover or make a tidy haystack in the corner of your goat shelter. Be aware that loosely stacked hay takes up a lot of space.

It will be difficult to cover all your goats' hay needs this way, but it is rewarding work (and a workout!), and homegrown hay most likely will be the goats' favorite meal.

Always introduce new foods slowly.

Other Staples of a Goat's Diet

Backyard goats with limited pasture and little access to browsing areas need a feed supplement along with their daily hay rations.

Pellet Food

Pellet food is a concentrated mixed feed which may also contain minerals, vitamins, and proteins. You need to choose a pellet feed that is right for your goats and for the geographic region you live in. The trace elements in feeds differ by region to supplement those that are missing from the grazing areas in a specific area. Feeds also contain different levels of protein, which will be clearly marked on the label. The quality of your goats' hay and their use will determine which protein level is appropriate. An experienced goat farmer, veterinarian, or agriculture extension agent can help you determine which feed to choose. Depending on what you are using your goat for and the quality of your hay, you may not need pellet foods at all. If you are milking your goats, you should give some pellets at milking time to help replace the nutrients you are taking from her.

When buying your goats, ask the seller where they buy their feed. Farm supply stores will normally carry animal feeds. Pellet food usually comes in 50- or 100-pound (22.7 or 45.3 kg) bags. If you feed an average of 1 pound (454 g) per day per doe, you will need 60 pounds (27.2 kg) per month for two goats—or 720 pounds (326.6 kg) a year. Pellet foods will vary in price, depending on quality and distance of delivery.

Pellet food should be stored in a large metal can with a closable lid that goats can't open. Large metal cans with two locking handles on the lids work well, though the best prevention is making sure your goats can't get out of their enclosure. Do not leave the feed in the bags within your goats' reach, as they will rip the bag open and overeat. (For a story on the perils of overeating, see "The Farmer's Mistakes" on page 69.) Mice and rats will also find your feeds unless they are stored properly. Do not feed moldy or otherwise soiled feeds to your goats.

Pellet food should be offered to the goats twice a day, once in the morning and once in the evening, in small portions to avoid overeating. The feeding instructions on the bag will indicate how much to give per feeding. An example might be "For milking does feed 1 pound (454 g) daily in two separate feedings" or "Feed 1 pound (454 g) daily per 100 pounds (45.3 kg) of goat." The latter example means if your goat weighs 125 pounds (56.7 kg), you would feed her 1.25 pounds (567 g) of pellet food.

Pellet Feeding Tips

- Like all new foods, pellet food should be introduced into your goat's diet slowly. When you bring your goats home, be sure to get the same food they were being fed in their previous home to avoid sudden changes.
- If you are just starting your goats out on a new pellet food, it's better to start with a low protein percentage and gradually work up to higher percentages should your goat need them.
- If you are changing from one pellet feed to another, start by slowly integrating 10 percent of the new feed in with 90 percent of the old. Two or three days later, you can up that to 20 percent new feed and 80 percent old feed, and so on. Never rush new foods into a goat's diet, as the risk to the rumen is very real.
- It's natural for us to want to give our goats the best we can find or afford, but be careful not to kill her with kindness. Feeding high-protein alfalfa with high-protein pellet food when the goat can't utilize it all is a waste of money and a possible health risk to your goat. Diets that are too high in protein may result in jumpy nervous behavior. Your goats may act like they're on a caffeine rush, which is not good for them. Worse, their rumen can be stressed, droppings may become runny, and goats may urinate a lot more than usual. (Also, high-protein feed and hay combined is very expensive!)
- If your hay is very high in protein, you may be able to keep your goats healthy with a low-protein pellet.
- You need to find the right balance of proteins to keep her healthy and productive without giving her more than her body can use.

What Else Can I Feed My Goats?

Can goats eat anything besides hay and pellets? Of course they can—and they will. If your goats have access to woods to browse, they will eat many different types of plants, both fresh and dead. The most important thing to remember when feeding goats is to introduce new foods gradually to ensure they don't disrupt the rumen.

If your goats will have access to grazing pasture, introduce them to it very slowly. In the morning, let them fill their bellies with hay before letting them out on pasture so they don't gorge themselves on fresh grass. At first, limit their time on grass to avoid overeating. Always let the morning dew burn off before releasing your goats to pasture so your goats aren't eating wet grass.

If your goat finds something to eat that she is not used to but she likes the taste, she will ingest large amounts of it very fast if the food is not introduced slowly. The sudden filling of the rumen with unfamiliar foodstuffs will send the bacteria in the rumen into overdrive, causing dangerous bloat.

Of course, occasionally, you or your friends will want to give the goats a treat. Keep it simple. A handful of cornflakes or raisins will have them following you like lovesick puppies. Be careful not to offer treats too often and keep them on a stable diet.

Keep mineral salt blocks available at all times. Hang them at the proper height—about the height of the goats' muzzles—so goats can reach them, but so the blocks do not get soiled.

Mineral Salt Blocks

Mineral salt blocks are just that, blocks of salt with traces of other vitamins and minerals that your goats can lick to help meet some of their daily dietary requirements. Think of it as a chewable vitamin. Salt blocks are usually regionally specific, designed to meet mineral and vitamin deficiencies for your general area. The manufacturers of mineral blocks combine the latest information on goat dietary needs and couple that knowledge with mineral studies of various geographical areas. For example, if an area has a general lack of selenium in pastureland, the mineral blocks will usually contain extra selenium to help make up the difference between daily needs and what a goat would take in under normal pasture conditions. The salt will also make the goat thirsty, encouraging water consumption for overall better health.

Caution: *Make sure you buy a mineral salt block that is made for goats. Not all blocks are okay for all animals.*

You may notice periods of higher and lower consumption of the salt blocks. How much salt a goat eats will depend on how balanced her diet is. If you see a lot of salt-block licking, it may indicate that her hay or pellet feed is not satisfying her nutritional needs. Occasional licking indicates that she is being provided for. If your goats don't lick at all, either they don't need it or it may smell dirty and need to be replaced.

Salt blocks come in a variety of sizes from less than 1 pound (454 g) to more than 50 pounds (22.7 kg). Buy a salt block that is appropriate to the size of your herd. If you buy one too big, your goats may not consume it fast enough, leaving the block to get dirty or rain damaged. A backyard herd of two goats requires only a small, 1-pound (454 g) block, replaced as necessary.

Salt blocks typically come with a hole through the middle for hanging. To hang a salt block, tie a stick to the end of a rope and slide the rope through the block, so that the stick holds the salt block in place. (Simply tying the rope around the salt block is dangerous because the goat could get her head stuck in the loop.) Hang the salt block where the goats can reach it but not lean or rub on it. You can also buy mounted wall boxes specifically for holding salt blocks or loose minerals.

Baking Soda

Baking soda—sodium bicarbonate—is an acidic neutralizer that helps stabilize the pH of a goat's rumen. It is the same baking soda you may have in your pantry and is available at any grocery store. Animal feed stores also offer sodium bicarbonate in larger boxes at a lesser cost. A small amount should be available to the goats at all times. Some goats will eat a little baking soda daily, and others will hardly notice it's available.

Goats in the wild don't need baking soda, as they naturally self-regulate their diets as they browse slowly, but baking soda can be helpful to goats in an enclosure, where they may eat large amounts of rich foods quickly.

Start by leaving out small heaps of the baking soda and watch the consumption. Goats who consume large amounts of baking soda over a period of time may need adjustments to their diet. Goats who take little nibbles occasionally are fine, and if your goats completely ignore the baking soda, congratulate yourself on feeding your goats a balanced diet! The soda should be fresh and changed often if not being consumed. Even though the goats may not eat it daily, sodium bicarbonate should always be available in case she does have sudden rumen upsets.

Make sure your goats have access to water at all times, and that the buckets are clean and full.

Water

Clean water should be provided at all times. The least time-consuming, most expensive way is to have an automatic water system. A float valve will keep the water level constant letting in new fresh water every time the goat drinks. If you live in a cold climate, consider a water system with a heating element to prevent freezing.

Buckets filled with water work just fine, although they will need to be checked more often. When it comes to water, it is always best to overestimate your goats' needs to be sure they never run out. If you find that the water level is often low, add another bucket. Make sure the buckets are cleaned regularly, as goats are particular about water and will not drink soiled water. Plastic buckets can develop a bacterial growth over time. To prevent excess soiling, keep the bucket on the outside of the goat pen and position it so the goats must stick their heads through a hole in the fence to access it. Clean buckets are important. If you wouldn't want to drink out of it, neither will she!

CHAPTER CHECKLIST

- ☐ Have you read—and reread—the section on rumination, and do you understand why it's so important not to overfeed your goats and to make feed changes slowly over time?
- ☐ Have you located a source for dry, good-smelling hay and arranged to have a year's supply available to you?
- ☐ Have you located a source for other goat feeds?
- ☐ Have you determined what feed-protein percentage is right for your goats depending on how you are using them, their access to pasture, and the quality of their hay?
- ☐ Have you closed the gate?

From the Farm: The Farmer's Mistakes

The first lesson every farmer learns is that as much as we want everything to be perfect for our goats, it won't always work out that way. You will make mistakes. You'll forget to shut a gate or mistakenly leave a bag of feed within reach. And those mistakes can have tragic results. I was reminded of this the first year at Valle di Mezzo.

I had ordered my normal three-ton (2.7 mt) delivery of food pellets for my milking herd. On this occasion, I had changed feed suppliers. The previous feed company trucked their feed in a large tank with an auger to unload the feed directly into my silo, conveniently stored out of reach of the goats. But the feed company that delivered this load brought the pellets in three one-ton (0.9 mt) nylon bags with no way of pouring the pellets into the silo. The delivery arrived on a rainy winter evening, and the tired driver unloaded the bags in my barn. I would deal with it the next day. I went about my nightly chores and mistakenly left the goats' gate open as I turned off the barn lights.

When I went down to the barn at my usual early hour the next morning to start my milking routine, I found my goats happily playing king of the mountain, climbing all over stacked hay. Goats were everywhere, except in their stall where they were supposed to be. I was frustrated, but the goats were happy, and apart from some poop to clean up and a delay in milking, I figured there was no harm done. But as I started toward the happy herd wondering how long it would take me to get fifty goats out of the hay, I discovered three goats lying on their sides, their stomachs grossly overextended. One of the new feed bags was ripped open.

I rushed to get the loose goats back into their stall and away from the food as the downed goats groaned in pain from bloat caused by overeating. Their sides were stretched tight like drums, their ever-expanding rumens suffocating them from the inside. The damage was done. I had arrived too late, and all three died within minutes of my first getting to the barn.

This is not a story I like to tell, but I think about it often—and then check again to see that my goats are safe.

CHAPTER 6

Getting Your Goat

In the first five chapters, you've learned all about goats. You've selected a breed. You've gotten the necessary municipal permits and talked with your neighbors. You've built safe housing and found a source of feed.

Now, it's time to get your goats.

You'll get lots of advice on buying your goats, but there is one rule you must never forget: never buy someone else's problem!

No matter what breed you've decided to keep, you are looking for happy, healthy goats. Observation is the key when you go goat shopping. Keep your eyes and ears open and use good common sense.

The goats should be lively, and rambunctious if young, and really seem to be enjoying life. The joy of raising goats is the communication and contact you will have with your animals, so look for goats that are friendly and trusting toward humans. Buying an unfriendly goat with the thought that you'll treat her with love and she'll "come around" could test your patience for a long time. Start with friendly goats and enjoy the journey.

Bottle feeding the kids will create a long-lasting bond between you and your goats.

County fairs and agriculture expos are great places to learn more about goats and meet fellow goat keepers.

Goat Sources

With the ever-increasing popularity of goats and goat products, you may not have to look as far and wide as you would have imagined to find a breeder near you. The Internet offers a wealth of information, and looking for goats for sale is no different. (See "Resources" on page 150 for some suggested websites.) There are also numerous magazines for goat farmers and backyard farmers, which often include advertisements for goats; these include *United Caprine News*, *Hobby Farms*, *Small Farm Today*, *Dairy Goat Journal*, *Homesteader* and *Acres*, and several others that are listed in "Resources."

If there is an agricultural or 4-H club near you, it can be a good resource for finding a local goat breeder. County, state, and other regional agricultural fairs with goat shows are another resource. Goat owners at fairs are usually very helpful and happy to talk about their animals and give advice. Goat shows are also a good place to see many breeds together, so you can compare their traits.

Auction barns also sell all kinds of farm animals to the highest bidder. The auctions are interesting and exciting, but they are not the best place to find a goat for your backyard. Auction barns are usually the end of the road for most of the animals marched through them.

Doe versus Doeling

Once you've decided how many goats to get (remember, at least two!) and which breed you prefer, you need to decide what age of goat you want to start off with. Will you purchase does or doelings? This decision is particularly important if you are raising goats for milk, but it is a consideration with any goat you plan to breed.

Choosing does:
If the main reason you want to raise goats is for the delicious milk, you may not want to wait a year or more for your kid to kid. You will be better off bringing home does already in milk. With does, you will have a good sense of what you are getting. The does will be near their full-grown size, experienced at kidding, and if you choose well, patient milkers and good producers. (Note: Milk production can decrease after a midlactation move. Once the does adjust to the new environment, milk production should pick back up.)

Choosing doelings:
If you start with doelings, you will have to wait for that first taste of fresh milk, but you will have the joy of watching them grow up. You'll become more comfortable with them without having the added chore of milking twice daily, at least to start. By the time your goats have grown to breeding size and gone through five months of pregnancy, you will know your goats well.

Doelings need extra attention before and during their first births but for the most part, things go smoothly. First-time milkers are not usually very productive—it often takes two or three pregnancies to achieve maximum production—but that can be a bonus for first-time goat farmers who don't want to be overwhelmed by milk immediately.

A beautiful, young, multicolored Nubian buck: Note that his ears extend beyond his nose.

What to Look For

At a basic level, shopping for goats is like shopping for anything. You wouldn't go to the grocery store and head straight for the rotten fruit bin or dented can section, and neither should you accept low quality when shopping for goats. Always look for the happiest, most curious, best-looking goat of the herd. When going to a new farm to shop for goats, your best ally is observation.

If the seller has already separated those goats that are for sale, compare the for-sale animals with the animals that are not for sale. Do you see a big difference? Do the for-sale animals seem less lively or less energetic? Do you have the gut feeling you are buying someone else's problems? Most goat farmers are honest and want to sell you the best goat they can in hopes of creating a repeat customer and good word-of-mouth, but as with all things, the rule is "buyer beware."

If the goats are in a far corner without much interest in what's going on in their surroundings, you may want to find out why. In addition to possibly being a sign of unfriendliness, this could be a sign of health issues. If you are just starting out with your first goats, it is imperative that you start with healthy animals. You want your first experience as a goat farmer to be enjoyable for both you and the goats. Don't bring home an animal you feel has some health concerns.

Most likely you are not a veterinarian, but you can look for physical warning signs. If red warning signals pop up at any time during your inspection of the goats, it is acceptable to ask a veterinarian to examine the goat you are interested in. Any money spent on a veterinarian's opinion will be more than earned back with peace of mind.

Observe the Goats

- **Feet:** Are they walking on all four feet without pain? Do the feet have an unpleasant smell? Both can indicate overgrown toenails or a foot infection. Healthy feet don't smell.

- **Weight:** Do they look really skinny? Underweight goats may suffer from worms or malnutrition.

- **Coat:** Are their coats shiny and healthy looking or dull and lackluster? A dull coat can indicate she is lacking nutritionally or needs a change in environment. More access to sunshine and fresh bedding can improve her condition.

- **Teats:** Do the goats have two teats? With both does and doelings, it's important that there are only two teats. Three or four teats are a defect that can be passed from mother to daughter, so these animals need to be culled. Although a three- or four-teated goat may otherwise be normal and healthy and her personality may be a perfect fit to your backyard, animals with abnormalities should not be allowed to multiply. The extra teat may even have milk to be milked out, may hang in the way of your milking, or may cause her kid to go hungry if it suckles on an extra teat that doesn't produce.

- **Horns:** Do the goats have horns? Horned goats are not recommended for the backyard farm for your safety and for the goats' safety.

When buying a goat, always check for a full set of teeth. Missing teeth or a misaligned jaw could lead to chewing and thus eating and production issues.

Examine the Goats

- **Diarrhea:** Lift their tails and check for signs of recent diarrhea. If their backside is dirty, ask the seller why. Diarrhea can be a sign of poor diet or worms, though occasional diarrhea may be normal.
- **Abscesses:** Feel for abscesses, which are lumps on the body that feel like a big pimple about to pop. Abscesses, depending on their cause, can spell trouble. Abscesses can form from insect bites, routine shots, and similar small punctures, which are not a concern because you know the source. Abscesses that have no explanation need to be examined closely. Some forms of abscesses are contagious and can infect the entire herd.
- **Herd health:** Look around the barn. Do you see many goats with abscesses? If so, quickly say your goodbyes and shop elsewhere.
- **Teeth:** Open their mouths and check their teeth. Are they all there? If a goat is missing teeth or has an undershot or overshot jaw, are they able to chew correctly? If they can't eat, they can't produce. *Note: Goats only have front teeth in their lower jaw.*

Horns and Their Function

Goat horns can grow to huge and beautiful proportions, sometimes spanning 3 feet (1 m) from tip to tip on a buck. Does have smaller, but still impressive, horns that can span 18 inches (45.7 cm). It's natural to assume that goats use their horns as a weapon to defend themselves against predators, but most often goats use their horns on other goats, which makes horns a hazard on the farm. (Their natural defense against predators is typically the ineffective practice of snorting and stamping their front feet.)

Polling and Dehorning

Horns have other uses. For instance, they provide temperature regulation (the blood cools as it circulates through the horns) and they allow goats to scratch their own backs. But their danger to you, other unsuspecting goats, and the goat herself, when those horns trap her head in the fence, outweighs these benefits.

Goats born without horns are said to be "naturally polled," but the majority of goats are born with little nubs that will develop into horns unless the goat is "dehorned" as soon as the buds start to develop, as young as three to ten days. The dehorning process—through burning or chemical removal of the horn bud—can be painful but will save your goat from the injuries that come with horns. (A visual overview of the debudding process is on page 77.) A third method of dehorning involves banding the horns tightly to cut off blood flow. This is both painful and ineffective and should be avoided.

If you are buying kid goats that have already been weaned, the horns will have grown to such a size that dehorning is not recommended. Especially if this is your first experience with goats, buying already-dehorned goats will save you the trouble, trauma, and money involved in this unpleasant but necessary task.

Horns, although beautiful and useful to goats (for back scratching, as a cooling system, and for fending off aggressors), should be avoided on backyard goats.

Goat Keepers' Glossary

"Get your goat": Goats used to be kept as stablemates for nervous and flighty racehorses to keep them calm. Owners could sabotage the race by stealing their competitor's goat, making the horse stressed and nervous. That's why when you see somebody looking nervous or confused, you may say, "What's the matter, somebody get your goat?"

Overview of the Debudding Process

Supplies

- 1 cc of tetanus antitoxin (to protect against tetanus)
- Debudding iron
- Anti-inflammatory pain reliever from your veterinarian. Aspirin or ibuprofen can be used if given with food.
- Antiseptic spray

Disbudding

A dehorning iron is used for disbudding young goats. The hair around the horn should be clipped or shaved prior to dehorning. This ensures a quicker process with less burning of hair and less smoke so you can see what you are doing. The iron is heated until red hot. While the kid is held firmly (for his safety and yours), the iron is placed on the horn bud and with rotating pressure around the circle of contact. Remove the iron and let it reheat before applying to the other horn. Pressure from the iron to the horn bud should be firm but not hard.

For proper disbudding, there should be a copper ring all around the horn bud. If you missed a spot, you can go back to it, but for less stress on both you and the goat do your best to make each burn count. Detailed instructions that come with the dehorning iron indicate how many seconds the iron should be held on the buds for best results. The kid will scream during the process, but it will seem to have completely forgotten about it seconds after being back with its mother or stable mates. Before you attempt this (or any) delicate procedure, it is a good idea to observe an experienced dehorner so you have a good idea of how to do it and what to expect.

With plenty of grazing space, a nice barn, and a raised platform for resting, this little kid must feel like the king of his mountain.

Who can resist spending time with the kids? Your love and care will help them transition to their new environment.

What to Ask the Seller

The number-one question to ask any goat seller is, "Why are you selling her?" The main reason farmers sell does or doelings is that the farm doesn't need all of the does born that year. Or perhaps the farmer's milk production plan has changed, requiring fewer females. Be wary of any answer that hints at a problem with the goats for sale.

Kids

If you are starting out with kids, it's best if the kids are already weaned and drinking water out of a pail. If you are just starting with goats and envision bottle-feeding your own kids so you can bond, know that bottle-feeding requires a lot of dedication and cleanliness, and kids on a bottle are still very vulnerable to disease. Do you have a source of milk three or four times a day for the next two to four weeks until you can wean her yourself?

Ask the following questions:

When was she born? Apart from putting her birthday on a social networking site, this is important to know if you are thinking about breeding her the first year. You don't want to breed until she is at least seven or eight months old.

Can you see the parents? Looking at the mother, you will get an idea of how your kid will look in a few years (unless she is a mixed breed). Look at the doe's udder system and think "like mother, like daughter." Would you want to be milking the mother?

Is she a single, twin, triplet, or quad? If she was a multiple birth, can you see her brothers or sisters? How does she compare in size? Is she the same or is she visibly smaller? If you want to breed your goat, it is important that she is of appropriate size. If your goat is going to be pet only, a smaller goat, or runt, is fine—and should be priced accordingly.

Does

Young does are usually energetic with smaller udders. Second-year does and does up to about five years old are elegant with big, firm, highly productive udders. Does older than six years will start to show their age. Their udders may droop, and they are often less energetic than their younger counterparts, but that doesn't mean they are no longer useful. With good care your favorite doe may live and produce milk ten years or longer.

Ask the following questions:

How old is she? Keep in mind that if you want to milk or breed her for a number of years to come, she may not be very productive after about year six.

How many times has she given birth? The average doe can give birth six times. The more times she has given birth before you purchase her, the closer to the end of her productive age she is.

Has she had mostly singles? Twins? Triplets? Does who often produce twins are more profitable for a farm, giving you more kids to milk or sell. However, you may not want a doe that has a history of triplets if your space and time are limited.

Does she have a history of difficult births? If the seller says yes, beware! Although any goat can have complications, you want a doe who can give birth with limited assistance.

Has she ever had mastitis? If the answer is yes, that's a big red flag. If she has had mastitis, an infection of the udder, in the past, she is at greater risk for more problems in the future. In the worst-case scenario, she will not be able to produce milk for her kids, leaving you to find a milk source for them. Avoid purchasing goats who have had a history of mastitis. (See chapter 11 for more information on mastitis.)

Is she milking now? If you want milk, find out how long she has been lactating. Most dairy goats give milk for eight to ten months after giving birth.

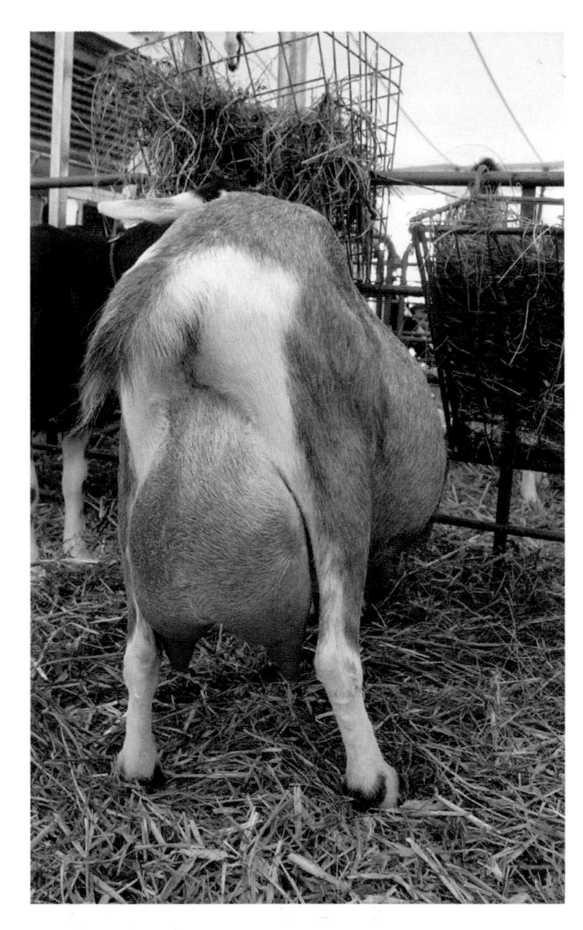

This is an example of a nicely shaped udder and teats for easy hand milking.

Milking Does

If you choose to buy a milking doe, you need to make sure her udder is in working order. Regardless of how much milk you hope to get daily, the goat's udder should be in good condition, lump free, and easy to milk.

First, observe the udder. Is the udder nicely compact and round with two easily milkable, downward-pointing, easy-to-grasp teats? Does she have problems walking because the udder is so big and bulky that it bangs against her legs? This means there is a risk of bruising and other damage to the udder. Is the udder high to the body (good), or does it hang like a heavy, water-filled balloon, skinny at the top and bulging at the bottom (bad)? Is one side of the udder bigger than the other, indicating that one half may be nonproductive? The sides of a healthy, well-formed udder should be equal, and both teats should produce the same amount of milk.

Second, touch the udder. A healthy udder should feel comforting, soft, smooth to the touch, without lumps of any kind. Some udders are easily pliable and seem liquid filled; others feel more meaty and firm. Both are acceptable. Squat down and really examine the udder with both hands. Don't be shy.

Gently palpate the udder with your palms and fingers. Does the udder feel hot? Are there blisters or lesions on the udder? Do you feel any lumps within the bag or hard spots that don't feel right? These could be signs of mastitis, past or present, or trauma to the udder. Turn the teats up toward you and examine the opening. Do they look red and sore? That is also a bad sign. If you plan to milk your doe, her udder needs to be in good working order.

Don't forget to observe the goat's demeanor. Is the goat stomping her feet and trying to get away from you when you touch her udder? That could mean trouble. It is a sign that she either doesn't like people, or at least strangers, or possibly something is irritating her udder and it could be painful for her. Hold the goat to do a proper udder exam and you may discover why she doesn't like being touched.

Try milking her. If the goat has passed the initial observation and examination and is friendly and willing to be handled, it's time to milk her. If you have never milked a goat, ask the seller to show you how. (Be prepared. Read chapter 10.)

Does the milk flow easily? Milk some into your palm and look at it. Does it look healthy and white or is it thick and off-colored with strings of blood? If the milk doesn't come easily or if it looks like anything other than milk, this is not the goat for you.

Taste the milk. If you are satisfied with your exam so far, ask if you can taste the milk. If you are buying from a milking or cheese-making farm, the animals will have been tested to be sure the milk is safe. Once that is established, milk a cupful and drink it. Do you like the taste? Is it better (or worse) than you thought it would be? Is it different but not unpleasant? You may not be used to fresh goat's milk, which many people love, but is it something you can get used to? All goat's milk, regardless of taste, will make fine cheese or soap bars, but if you plan on drinking it every morning, be sure you like the taste of the milk before you get your goat home.

Very important: Be sure you like the taste of your new goat's milk before you bring your goat home.

What to Pay

Goat pricing is based on perceived quality. You can pay a lot of money for a healthy goat that has a good pedigree (traceable heritage), high milk production, and is a classy looker, or you can find better prices for a healthy goat that may not produce as much in the pail or without the pedigree. Either way, make sure you get a healthy goat.

If you are looking for a kid for your kid to show at a goat show or if you plan to sell your goats' kids as future producers, you'll want to invest in an animal that comes from a proven line of productive goats. But if you just want fresh milk for home consumption, you don't need a highly pedigreed (and highly priced) show goat.

One of the many perks of raising your own goats, early bonding lays a foundation of trust with your goats, which leads to easier management when they are grown.

Here are two examples of transportation options: Open-bed pick-up trucks work fine for short drives. Make sure the goats are tied securely to prevent them from jumping over the sides. A dog crate is perfect for transporting small kids.

Transporting Your Goats

If you are starting out with kid goats, the easiest way to bring them from the farm to your home is in a dog crate. An eight-week-old kid goat weighs about 35 pounds (15.9 kg) and will fit in a medium-size dog crate. If you are buying two goats, choose a crate in which they can fit snugly together and help calm each other. After a few minutes in the crate, they will usually lie down, quiet down, and enjoy the ride to their new home. If you are putting the crate in the backseat of your car, be sure to put down plastic or an old blanket to protect your seats from nervous pee.

When buying full-size goats, the transportation issues are a bit more challenging. If you have access to a small livestock trailer or pickup truck with a closed cab, those are the safest and least messy ways to travel with your goats. A less-ideal option is to take your goats home in the open bed of a pickup truck.

When transporting your goats in an open cab, you must follow these three safety rules:

- Tie them very closely to the cab.
- Make sure the rope is short enough that they can't go over the side, but long enough that they have sure footing and won't slip and choke.
- Obviously, you need to keep an eye on the goats throughout the ride.

Most often, though, backyard goats ride home in the backseat of the car, securely tied to a short rope, for safety reasons. You don't want them jumping around while you're driving. Cover the backseat well with plastic. You are guaranteed a big cleanup job by the time you get home.

Alternatively, just ask the seller if they will deliver!

CHAPTER CHECKLIST

☐ Have you decided how many goats of what breeds and ages you will purchase?

☐ Have you located a reputable goat source?

☐ Do you know what to look for when observing goats for purchase? Do you know what to ask when buying goats?

☐ If you are buying milking does, have you milked the goats and, if possible, tasted the milk?

☐ Have you considered how you will transport your goats to their new home?

From the Farm: The Road to Priello

At Priello I have a truck and small livestock trailer to transport my animals, but it that's not how I started.

The road up to Priello is long, and it climbs steeply. The first few years, I rarely had access to a car, and even when I did, the dirt road wasn't always navigable. So first, when I bought five chickens, I walked them up the mountain in a big cardboard box. It must not have been as uncomfortable for the chickens as it was for me; when I got home I found that one of them had laid an egg along the way.

Next came the turkeys. The four turkeylings traveled in a bigger, heavier box. About a third of the way up the mountain, it started to pour. I was soaked. Worse, the turkeys were soaked. Turkeylings are notorious for being *delicatissima*. One of them didn't make it, but the other three lived to see brighter days.

And then there were the guinea fowl. I had taken a bus into a nearby town, Anghiari, to attend the Wednesday farmers' market, where I bought guinea fowl. I planned to take the bus home and walk up the mountain as I had with the chickens and the turkeys. But there were no return buses running on Wednesdays. I tried to hitchhike, but no one stopped. I walked the entire seven miles (11 km) home with a box of four screaming guinea fowl.

The horse was the worst. After dreaming for years of getting a horse, I found Chérie—a tall and proud dark bay mare living in abandoned conditions down the mountain. She hadn't been handled in some time or ridden in years, so I walked her home, a three-hour trip, trying to keep her calm by repeating her name in my best French accent. At Priello, I gave her a good brushing, trimmed her mane, tail, and hooves, and set her out in the pasture to make friends with our other animals. After a couple of happy leaps, she fell over dead from the excitement. She had been on the farm less than an hour.

After all that you would think that I would have learned my lesson by the time I got my first goats—but my friend Alessandro would disagree. We once brought a buck home at the height of rut in his brand-new car. I get carsick in the mountains, so I drove, and Alessandro had to hold the foul-smelling beast in the backseat with all the windows open.

CHAPTER 7

Basic Goat Care

Congratulations. You have goats. Now what? The first time you bring goats into your life is very exciting for both you and the goats. You are thinking of fresh milk and cheese and looking forward to the peaceful morning chores. Your goats are exploring their new home, sniffing and snorting, sometimes even kicking up their heels in delight.

Be sure you are properly prepared for your goats—and be sure to have some hay waiting for them in their feeder. With the excitement of travel and a new house, your goats will take comfort in grabbing mouthfuls of hay and calming themselves as they study their surroundings. You should study their new surroundings, too. Is there anything you may need to change in the enclosure? Is there anything you overlooked? Ask yourself: are the goats taller than you imagined? Can they reach things you thought would have been out of reach? Are the goats already smarter than you imagined? Are the gates closed in such a way that a goat can't open them?

Once you have ensured the goats are safe, it's time to get to know each other.

Kids at play will keep even the most seasoned goat farmer entertained.

A small treat can be a big step in building trust with your new goat. A little scoop of their regular pellets will do the trick. They will look to you as the food source and even the shyest goat will be won over.

Introduce Yourself

Nothing wins over a goat like food. When introducing goats to a new home, you want to make them feel as comfortable and welcome as you can as quickly as possible. The faster they get into the routine and feel part of the family, the better.

- Let them nibble a few raisins out of your hand as they sniff your clothing and probably chew your pant legs. They'll learn to come to you looking for raisins and soon enough settle just for a little petting. (As always, easy on the treats to prevent digestion problems.)
- Talk calmly to your goats while you go about your chores. Let them learn your voice.
- Give little pats and get them used to your touch slowly.

The first few days of goat ownership are for getting used to each other and making any necessary adjustments to your image of goat farming so that this experience is a happy one for all involved.

Observe Your Goats

The most important tip for a first-time goat farmer is the same as the most important tip for a first-time goat buyer: observe the goats. Knowing how your goats typically act will help you notice, diagnose, and solve problems quickly, keeping your goats healthy and happy.

A Happy Goat

Most goats quickly settle into their new routine, and it will seem as if they have been in your goat house from day one. If your goats look happy and alert and they are alternating between periods of eating and periods of resting and chewing cud, their transition to their new home has been a success.

This doe seems happy with her housing.

To ease the transition, be sure to keep the goats on the same feed and feeding schedule they had at their previous home. This will keep their rumens working smoothly, and the familiarity of mealtimes will keep the goats comforted while they learn their new barn.

An Unhappy Goat

But what if your goats do not appear to be adjusting well? An unhappy goat may show signs of withdrawing into herself with little interest in exploring her new world. She may eat less and stand with hunched shoulders. She may just look sad, and perhaps she is. Remember you have taken her away from what she knew. Be patient and give her a week or so to learn her new routine. Be kind and gentle and keep your schedule so she'll know what to expect.

If you have brought home two goats, which is strongly recommended, your goats are likely to adjust more quickly. If you are introducing a new goat into a preexisting herd, it may take longer. Remember the herd hierarchy. If you already have a willful, confident goat at home, your new goat may be bullied.

If your goat is healthy but still unhappy, consider what you can do to improve her living situation. Does she need more space? Closer contact with other goats? Less contact with other goats? More attention from you? Less attention from you? If you observe your goats closely, they will let you know what they need.

Goats are happy by nature, and chances are that yours will be too. If you think her adjustment period is taking abnormally long and she doesn't act the way you think a healthy, happy goat should act, do not hesitate to call a vet and make sure there isn't something physically wrong with her. If this is your first goat, you can't be expected to know every sign she is giving you, so start off on the right foot, and if you sense something is not right, call a vet or trusted goat-keeping friend and ask advice.

Know the Poop

Poop, feces, manure, droppings, dung, pellets, berries, or doe doo, whatever you want to call it, goat stool is a good indicator of the overall well-being of your goats. When a goat is healthy and the rumen is working properly, the stool will be dark green, firm, round, and uniform in size, and consistently expelled throughout the day. If you monitor what kind of droppings your goat normally produces, you will be able to quickly see any changes. Changes in stool are a warning sign of illness or nutritional issues.

If the goat berries become larger and moister, the goat is most likely ridding itself of toxins or bad bacteria. Continue to watch your goat closely for further changes.

Runny stool is more of a concern. If your goat is also acting differently, take a rectal thermometer reading. (See chapter 11.) If the reading is above normal (102°F to 103°F, or 38.9°C to 39.4°C), call a veterinarian. A veterinarian should also be consulted if runny stool is accompanied by weight loss, which can indicate that your goat has worms. A veterinarian can diagnose and treat common parasites.

Most often, runny stool that is not accompanied by a change in behavior or a fever is a side effect of a change in diet. For example, if your goat suddenly has access to a full day on pasture, runny stool may result. Introduce access to the pasture slowly. If you changed feeds, change back and see if the problem clears up.

Any change in goat droppings should be monitored. If you notice a consistent problem, call your veterinarian.

Train Your Goats

Whether it's working nine to five, church on Sundays, or news at eleven, everyone appreciates a routine. Goats, too, are animals of habit. Your goats will perform better and be healthier if you can put them on and maintain a schedule and routine.

The schedule: Consistency matters more than what the clock says. You can feed your goats dinner at 6 p.m. or 8 p.m.—as long as you do it at the same time each day. A schedule is particularly important for milking. Your goats' internal clocks will tell them that it's milking time, and their udders will be filled with milk.

If for some reason you need to change the regular feeding and milking times, do so slowly. Shift the schedule ten minutes each day until you eventually reach your new scheduled times.

The routine: Working with goats who know the routine doesn't feel like work at all. Just as goats appreciate and perform better with a consistent schedule, goats will, in time and with positive reinforcement, learn and follow a consistent routine. It's impressive to watch, and it's immensely satisfying to be so in sync with your animals. For instance, if you always say the same things when you prepare to milk, the does will respond to those vocal cues and come to the milking area willingly. On the other hand, if you are normally a quiet person, but decide one morning to play loud music during milking, the goats may be distracted and confused, uncertain what to do. Keep your routine, whatever it is, and your goats will know what's expected of them and what they can expect from you. If you need to change your routine, as with a schedule, introduce change slowly and consistently until she learns.

A herd of Saanens and their kids enjoy the green pastures of early spring.

Calling Your Goat

Goats can learn their names and can do so rather quickly, as long as you choose goat names that are distinct from one another. Give each goat some special attention while saying her name over and over and she'll learn to turn her head and look at you when she hears it. Eventually she'll come over to you when you call her. Goats are herd animals and like a leader, so in time you'll be able to call her and she'll follow you around the pen. This isn't just a cute trick. It comes in very handy if for some reason your goats escape and you need to get them back in their pen.

This is an example of what not to do while leash training: engage in a tug-of-war with your recalcitrant goat.

Walking Your Goats on a Leash

Leash training is important even if you don't plan to walk your goat like a dog. Simply, the more manageable your goat is, the better. If your goat ever needs to be walked anywhere, you will be thankful you put in the training. Walking even the shortest distances with a goat who doesn't want to cooperate will test your patience and sanity.

If they need something, even if it's just attention, goats can be very loud when they want to be.

- A sturdy leather collar that will not slip over the goat's head works best. It's never a good idea to leave collars on your goats full time, as they may get caught up in fences and hay feeders. Goat supply or farm supply stores will often carry "break-away" collars for goat safety. These are collars that fit around your goat's neck and will break away if she snags herself on something.
- It will be easiest to train a young kid, especially one that has been bottle fed and is accustomed to humans, but most goats will learn to walk on a leash. Although a small treat like raisins can be used to reward good behavior, do so sparingly. Goats can, of course, be trained without any food incentive—just substitute a good rub or petting to reward her progress.
- To start leash training, attach the leash to your goat's collar and give the leash plenty of slack. Don't expect your goat to walk with you right away. She needs to move around and test the gentle restriction of the leash.
- Once your goat is comfortable with the leash, practice walking on the leash with short trips around the yard until your goat trusts you and is happy to go where you lead.
- Be careful not to overdo a training session. At the beginning, your goat's attention span will be short. Set easy, short objectives and practice each day until you both feel comfortable before venturing out of the backyard. If the goat has a bad experience during your first outing because she wasn't ready, she is not going to want to go again.

Once properly leash trained, goats are almost as easy to walk as dogs, but don't get the idea that your goat will go jogging with you! Goats like a leader so will most likely follow you with very little leash pressure.

Curbing Bleating

Bleating is how goats communicate. It can be loud if they want something. Sometimes they just want to talk. They may bleat at you from across the yard just to get your attention. Charming as that is, your neighbors may not be so impressed. If a goat has all her necessities within reach and is accustomed to a schedule and routine, she is less likely to bleat for your attention. Changes in schedule and routine are the most likely cause of frequent bleating.

Tip

Edible Landscaping

When you do take your goats out for walks, be respectful of neighborhood or public plantings. The two of you will probably have two very different reasons for wanting to go for a walk: you will want the exercise and she will want to eat every plant she can get her lips on. Be mindful of the decorative plantings, as many of them are poisonous to goats.

Goat Grooming and Maintenance

A healthy goat is a happy goat, and to keep looking and feeling her best, she needs some routine maintenance.

Exercise

When keeping goats in a small space, exercise is important to avoid overweight goats. Some daily exercise time is ideal.

Goats are good hikers. Walking your goats burns up their excess energy and calories, aids digestion, and keeps their joints working properly. Plus, your goats will grow to love it!

Young goats, aged six months and up to about two years, in normal physical condition, can walk a mile (1.6 km) or more easily. Goats are not distance runners; they like to walk, and you will have to make frequent rest stops. It's best to start on short walks to see how your goat reacts to leaving the house and things familiar to her. As your does age and become more productive, they will slow down and may not want to walk as far as a growing kid. Older goats with bigger udders will need more resting periods and will pant if walking too far or too fast. Stop and let them rest. Goats are sensitive to heat, so avoid long walks on hot summer days.

Having something for them to climb or jump on inside their pen will also provide some exercise, fight boredom, and probably provide their favorite resting spot. A stump or large rock or picnic table they can jump on and off will help them stay happy and busy.

Brushing

Goats, like most of us, enjoy a good massage. Giving them an occasional brushing with a sturdy bristled brush, such as a horse brush, will keep their coats looking great and will give you an opportunity to bond with your goats. Consider attaching the brush head of a big push broom to a wall that your goats can rub against whenever they have an itch.

These young Oberhasli will get plenty of exercise playing on this simple cable spool, repurposed as a platform.

Hoof Care

Depending on your climate, your feeding routine, and the exercise the goats are allowed, you will have to trim their hooves once every six to twelve weeks. Goats with a rich diet have fast-growing hooves, and those who live on soft bedding have no way to naturally wear down their hooves (compare this to a goat who climbs on rocks all day).

The outer part of the goat hoof can be likened to a human fingernail and is continually growing. The excess must be cut away to remain flush with the bottom of the foot. If we let the hoof continue to grow, it will curve under the foot, where it will trap bedding material, feces, and dust against the foot. This can lead to lameness, weak hooves, and exaggerated foot extension, a painful condition where the foot seems to be pointing up.

How to Trim the Hoof

Each hoof is divided into two parts, almost like two toes. Stand or squat in a comfortable position beside the goat [a]. Pick up a foot and gently clean away any debris from around and under the hoof [b]. With your pruning shears, start at the back of the hoof and trim away the excess in little snips in a straight line until you have cut from the heel to the point of the toe [c and d]. Next trim the inside of that toe. Repeat with the other half of the hoof.

For your first hoof trimming, it is best to have an experienced goat farmer or veterinarian show you how to properly trim the hooves.

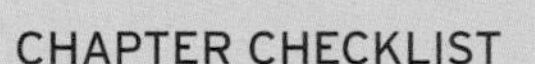

CHAPTER CHECKLIST

- ☐ Have you developed a habit of observing your goats?
- ☐ Have you established a schedule and routine?
- ☐ Have you trained goats to come when called and walk on a leash?
- ☐ Have you incorporated exercise into your goats' day?
- ☐ Do you have a supply of raisins, just in case?

To trim hooves, grab the goat foot firmly and always cut away from you.

Before cutting, remove any debris trapped under the overgrown hoof wall.

Trim both sides of the hoof starting at the "heel" and working toward the toe.

Trim off any extra growth on the tip.

From the Farm: My Garden of Eden

I have close to fifty Oberhasli goats in my barn, and even though they are basically the same color—red and black—I can tell who's who by their habits, how they eat, where they sleep, and how close they stand to me when we are out walking.

I milk at 5:30 a.m. and 5:30 p.m. Each time I am greeted with a chorus of bleating as the goats line up to come in. I use a mechanical milking system that milks twelve goats at a time. Each day I am almost guaranteed that P.U., a particular favorite of mine, will come in with the fourth group and she will pick the first milking position. I know Valentina will always be in the second group, and she likes to stand in the next-to-last slot.

This routine is helpful for me and for the goats. Milking is easier, but knowing how each goat acts and reacts is a great tool for avoiding health problems.

I once read about a lady who planted a "Garden of Eden" full of medicinal plants. Whenever she noticed one of her goats acting out of sorts, she allowed that goat time in the medicinal garden to eat what she needed to self-regulate. I loved the idea, so I made a similar garden. I liked putting the goats in the garden and watching what they chose. Soon they had eaten everything, and nothing grew back! Now I use the woods surrounding my farm as my big Garden of Eden.

On our daily walks, the goats and I usually head to a lightly wooded, grassy access road. The goats by now are used to my every move, so I can sit in the shade and they will slowly spread out, never venturing out of sight, eating whatever they feel like. There is a lot of tree-damaging ivy, and my goats do their best to eat it all. From there we move into the abandoned olive grove, which is overrun with wild, spiny blackberries. After they've eaten their fill of brambles, the goats seem to want grass, so we spend about twenty minutes on lush grass before moving toward the creek where the goats feast on the plants at the water's edge. Finally, we walk along a road toward home, the goats nibbling on pine needles all the way.

I keep an eye on what they are eating on different days, so I can better understand their behavior and cravings when they are in the barn. In this way, the goats tell me what they need.

CHAPTER 8

Breeding

As autumn approaches, it's time to think about breeding your goats.

Although some goats—like the popular Nigerian Dwarf and Pygmy breeds—can breed year round, the majority of goats breed in the fall. There is natural logic to this cycle: goats that breed in the fall will give birth to kids in the spring. The newborns have six months of good weather and plentiful food before facing the harsh, lean winter.

The average goat gestation is 150 days—about five months—and it is important for the new goat farmer to know that goats frequently give birth to twins or triplets. For the backyard goat farmer, that means a herd can triple or quadruple in a single season.

Breeding season means a host of new responsibilities. You will have preparations to make for breeding and birthing, and you will need to spend more time observing your goats through their heat cycles and pregnancies.

The most important question to ask as breeding season approaches is "Do I want to breed my goats this season?" If your goats are healthy and you are enjoying the responsibilities of being a goat farmer, breeding is an option.

Toggenburg dairy goats

The kids patiently wait for their breakfast.

Buckets with multiple nipples make easy work of feeding a large numbers of kids.

Deciding to Breed

There are two reasons to breed your goats: for the milk or for the kids.

Milk

If your main reason for keeping goat is because you want fresh milk, sooner or later you are going to have to breed her. Although you can buy a doe already in milk, by the time fall comes around she will most likely start drying up. It's time to breed!

Growing Your Herd

Of course, breeding means kids. If you have bred two does, you should expect between three and five new kids every year.

About half of the kids will be male and half will be female. If you have the space and the urge to increase your herd, for health purposes it's better to grow your herd from within. Raising your own doelings, although time-consuming, will lessen the risk of outside disease being introduced into your herd.

Kids to Sell

If you don't want to grow your herd or your doe gives birth to bucklings, you can sell the kids—and recoup some of the money you have spent over the last year. Try to find homes for the doelings if they are of good producing stock. Look for markets for the kids that will be sold for meat.

Deciding Not to Breed

There are also many good reasons not to breed your goats. If you are keeping your goats mainly as pets, you may not want the added responsibilities that come with breeding and birthing and later, kids and daily milking.

Space

Does your current housing setup allow you to section off some space for a private birthing area? If you want to keep the kids, will your goat shelter and exercise areas accommodate more goats than what you already have? If you don't have the space, don't breed your does. Remember that cute kid will be a full-size energetic goat in one year.

Time

Having a couple of pet goats in the yard is far less time-consuming than raising kids and milking mothers. During breeding and birthing, your schedule will be dictated by the goat's schedule. (For instance, you may need to call out of work when your goat goes into labor.) If you plan on raising the kids on a bottle, your time commitments for the first four weeks, or longer, will be both enjoyable and exhausting, as you have twice-daily milkings and a minimum of four daily bottle feedings.

The Breeding Process

The breeding process is both simple—buck meets doe—and complex. To breed your doe, you will need to locate a buck, wait until your doe is in season, get her to the buck, and hope they will mate.

Finding a Buck

If you decide to breed your goats, you must find a buck. Arrangements should be made ahead of time. Most breedings occur between the end of August and mid-January in the Northern Hemisphere and February to mid-May in the Southern Hemisphere. Call your buck owner at least six weeks in advance.

- When you buy your goats, ask if the seller has a buck for breeding or has recommendations. If the seller cannot recommend someone, call local farms, search the Internet and perhaps, place ads in local farming supply stores until you find a match.
- If you have purebred does and want purebred kids, you will need a buck of the same breed. If you are interested in kids no matter what their pedigree, your hunt for a buck of any breed will be much easier.
- When you find and make an agreement with a buck owner, discuss a plan of action. This plan will include days and times when you can bring your does to be bred. This plan is flexible because your does must be in heat.
- Keep the buck owner informed of when you are coming as a courtesy to the farmer's—and the buck's—schedule. Breeding season is very demanding on a buck.
- The buck owner has costs for keeping a buck, so be prepared to pay a nominal fee or barter for the services.
- Keep in mind if you bought goat kids when they were small and easily transported, your now-breeding-size doe will be a little trickier to get in the car, especially if you are taking two for breeding. Have your transportation figured out before you need it.

Rut

When mating season arrives, the first sign is a smelly buck.

When a buck is in rut, he will give off an odor that you won't soon forget. If the buck is kept in the herd with the does, he will most likely spend his days sniffing and chasing around the females as they pee to see if they are in heat. He will stretch his neck out straight, tilt up his head, curl his upper lip to expose his gums, and stand this way a few seconds as another way of detecting heat. If he does detect a female in heat, he will not let her rest, following her around and chasing her through the herd until she will stand for him.

Heat

A doe's heat cycle varies from eighteen to twenty-one days and lasts approximately twenty-four hours per cycle. If a doe is not bred, she will continue this cycle until midwinter, when the cycle naturally stops. A doe can only be bred during heat, and within those twenty-four hours, there is a limited window in which the does will be very interested in the buck. This is known as standing heat, when she will stand for the buck and allow him to mount her. You need to recognize her heat signs so that you can deliver her to the buck quickly. Again, observation is key. If you know how your doe acts normally, you will be able to detect her heat cycle just by looking at her.

The most obvious signs of heat are the following:

- Tail wagging and lots of it
- Increased bleating and a swelling of the genitals
- One of your does may mount the other one as a sign of heat. (Either one could be in heat.)

A Saanen buck with majestic horns and a long, flowing beard keeps his eye on the does in a far pasture.

To be sure that your doe is in heat, use a buck rag. A buck rag is a fabric rag that you will never use for anything else ever again. You must find a buck in rut, the smellier the better. Rub your buck rag all over the buck's head, chin, beard, wherever he smells. Quickly put the rag in an airtight jar. (A canning jar with a secure lid works great.) Keep the jar closed to keep it smelling strong. It's better to keep the jar out of smelling distance of the does. When you start to notice your does acting as though they might be in heat, take the jar out to your goats and open the lid in front of them. If they are in heat, you will know immediately. They will be very interested in the jar's contents with accompanying tail wagging and bleating.

Your does may be in heat at different times. Any doe that is not in heat will show only passing interest in the buck rag. For best results, use the jar sparingly so the goats don't become too used to it.

If you open the jar and your goats are interested, get on the phone to your buck owner and tell him you are on your way.

Breeding

The actual breeding is the easy part. After the checks and controls and constant observation to make sure the doe is in heat and getting her to the buck, once there if she is in standing heat the entire union will last about five seconds and you'll be left wondering, "That's it?"

Your does have been living in your quiet backyard setting for months. Getting in the car, riding over to a new barn, and being suddenly subjected to a barn full of goats with the sights and sounds and smells she hasn't experienced in a while may be overwhelming to her. In heat or not, the doe may be too distracted to stand still for the buck immediately. Let her get used to the surroundings.

The buck will do his courting routine, blubbering and pawing and chasing the doe around. The doe will likely settle down. Otherwise you can leash the doe.

They may mate three or four times or more in the space of ten minutes. The actual breeding happens so fast, you may think it might not have worked. If the buck has made a connection, your doe will suddenly hunch her hind legs under her belly like she just got goosed. Congratulations—your doe is now most likely pregnant. Keep record of her breeding date and keep an eye on her when she would have cycled next. Check for signs of heat or let her smell the buck rag. If she is excited again, she may not be pregnant and will need to be bred again.

A buck displays the curious upper lip roll as part of his courting ritual.

Goat Keepers' Glossary

Standing heat: This is the time period within the heat cycle that the doe will stand and allow the buck to mount her.

Pregnancy

Your doe will be pregnant for five months, but she probably won't start to show until well into the third month. Likewise, you likely won't see a behavioral change in your goat until late in the pregnancy. If your doe was milking prior to mating, she will keep milking for about two months. By month three of her pregnancy her milk will slow down, and she should be dried off to put her energies into the coming births. (See chapter 10 for more on milking.) First-time mothers will show physical changes sooner, as their udders start to form but not fill.

Feeding

Once your pregnant doe is dry, she will do fine on good-quality hay alone plus any amount of forage she is provided. Your goat should always look healthy but not overweight. During the last month of pregnancy it's best to give her small rations of pellet every day to help her meet the nutritional needs of the kids growing inside her, but you don't want to give her so much that she produces large kids, which will be a strain on her. Ask your goat feed source if there is a special prenatal mix for goats.

Exercise

Exercise is good for everyone, your pregnant goat included. The end of pregnancy is not a time for her to be lying around getting fat. If your leash training has been successful, try to get your does out for a walk every day. Walking will exercise her joints, stimulate her appetite, burn off excess energy, and also give her peace of mind. Most importantly, goats given room to move about seem to have fewer problems in birthing, possibly because the kids are moved into or kept in the correct positions prior to birth.

A beautiful, pregnant Angora doe waits patiently while tethered by a leash.

CHAPTER CHECKLIST

- ☐ Have you weighed the pluses against the minuses of breeding your goats?
- ☐ Have you decided what you will do with the kids?
- ☐ Have you found an appropriate buck and made arrangements with the owner?
- ☐ Have you been observing your goat and do you recognize signs that she may be in heat?
- ☐ Have you figured out how to get your goats to the buck?

From the Farm: Looking for Love

For the goat farmer, breeding season is all about logistics: where will you find a buck? When is your doe in heat? How will you transport the doe to the buck? If you make a living from your farm, the season is even more fraught. Kids are your livelihood. Will it be a good season like the most recent one that bore eighty healthy kids or a disaster like the year my ram, Rambo, had an undiagnosed testicular issue, leaving me with zero spring lambs and no milk for my cheese? As the great modern homesteader Carla Emory wrote, "Your animals are sure to multiply with joyful abandon . . . providing they have no economic value." (Evidence: The many cats running around my farm.)

But for the goats and other farm animals, breeding season is not about logistics or economics. For them, breeding season is all about instinct. Given a little bit of freedom, the animals will act on that instinct.

Thirty seven thousand acres of woodland starts just a stone's throw from my barn door. Every day, the cows, horses, and sheep—but not my beloved goats, who would have destroyed the orchard if I had let them wander—would go up into the mountains to graze on abandoned meadows and thick woods. The sheep and cows came home every night. A simple yell or a yodel and they would return obediently to the farm for milking and feeding. The horses, Pepi, Janka, and Lil'Baby Sally, were more independent. The mares would be gone hours, sometimes a day or more, but I had no reason to worry. They always came back eventually.

Until one autumn—I hadn't seen the horses for several days and was just starting to think "uh-oh," when the phone rang. It was the *carabinieri*, the military police of Caprese Michelangelo. "Ciao, Brent," the officer said. "How are you? How are your horses?" I knew I was in trouble.

My horses were on TV news. They were in the lost and found in a town two valleys away! Pepi, a beautiful muscular palomino quarter horse, had gone into heat and decided she was going to take matters into her own hands. She went looking for love—and had wandered 11 miles (17.7 km) over the mountain with the other two tagging along behind her. When I arrived to claim the wayward horses, the carabinieri gave me a lecture about keeping my animals under control and sent me and the horses home. Pepi (and I) did get some exercise, but she did not find a mate!

CHAPTER 9

Birthing

Few things are as exciting in the goat year as birthing of new babies: Will today be the day? And few things are as nerve-racking for the new goat farmer. The secret to enjoying the anticipation of birthing season is preparation.

The first step is writing your veterinarian's phone number in large, easy-to-read numbers within sight of your pregnant goats' pen. The second step is remembering that fewer than 10 percent of goat births need human assistance. In the majority of births, the kid or kids will arrive and you'll wonder what all the fuss was about.

Understanding the normal birthing process will help you recognize the warning signs if your goat needs assistance. If complications arise, it is most likely a kid in an abnormal birthing position. As long as you or your veterinarian acts quickly and confidently to realign the kid into the correct position, a successful birth is likely. Reaching into the goat to reposition the kid can be messy or unappealing, but knowing what you are feeling for and knowing that you are helping your doe makes the task much easier and natural.

But in most births, the goat farmer's only job is to stand back and watch the miracle.

Goats are known for being good mothers—very protective and attentive to their kids.

The Farmer's Preparations

You've marked due dates on your calendar. One hundred fifty days from breeding, plus or minus five days. Use this time to prepare for the kids' arrival. Being as prepared as you can be will help your confidence level when the time comes.

Contact your veterinarian. As a consideration, let your vet know that your doe is nearing her due date. This can be a busy season for vets. Do you have a backup number to call if your vet is on vacation or on another farm visit?

Arrange the birthing space. If your goats are friendly and loving to each other, they may be able to give birth in the same pen, but if you suspect that one may bully the other, have a separate birthing pen ready and put the expectant mother in there four to five days before her due date so she can get comfortable with her private nest. Some goats may enjoy the privacy of having a box all to themselves, others may be more stressed due to separation from herd mates. Use good judgment. Bullying in the goat world is normal, but try not to let it happen during kidding.

Arrange the birthing boxes for the kids. A doe and her newborn kids do not need a lot of space. A minimal space of 5 feet by 5 feet (1.5 by 1.5 m) is enough for them for the first few days while they get to know each other. Newborn kids can contract infections from soiled bedding through their umbilical cords, so keep plenty of fresh straw in their box. If you are separating the kids from the doe at birth, have the kid box ready with fresh bedding and heat lamps if your climate calls for it. (If you are using a heat lamp for warming, make sure the kids have room to move away from the heat source should they become too hot.) Kids will not need much space for the first weeks. Most of their time will be spent sleeping and bottle-feeding. Newborn kids will pee a lot, so keep plenty of fresh straw on hand to keep them dry.

Collect the items for your "baby kit." This kit will include all of the basic things you may need in a typical birth:

- Nonscented, antibacterial soap for you—should you need to assist with the birth, wash your hands thoroughly with the soap. This can also be used as a lubricant for your hand.
- Rubber gloves—Rubber gloves will help cut down the risk of infection from you to your doe should you need to assist with the birth.
- Towels—Both for you and for your kids
- Bucket for warm water for cleaning
- Iodine for dipping the newborn kids' umbilical cords and feet to prevent infection
- Thin string or dental floss to tie off the umbilical cord.
- Hair dryer—Depending on your climate, it might be very cold when spring babies are born. A hair dryer will act more like a heater to keep the kids—and you—warm.
- Baby bottle—A baby bottle for a human baby is a bit too delicate, and the nipple opening is too tiny. Buy a supply of nipples and bottles specifically for goats.

Together or Apart?

As your doe prepares to give birth, you have an important decision to make: will you leave the kid with his or her mother or will you separate the two immediately after birth?

There are many arguments for separating the kid from the mother at birth. Some of the most common are that you need steady milk production for your own consumption, you are concerned with the wear and tear of a kid feeding at the mother's teat, or you are concerned about CAE, a debilitating disease passed on from doe to kids. (See chapter 11 for more details.)

Arguments for keeping the kid with the doe are just as strong. It's unnatural to take kids from the mother. And it is a lot of additional work for the farmer. In addition to milking the doe, you will need to bottle feed the kid several times a day for eight weeks or longer. (On the other side of the argument, bottle feeding produces much friendlier kids that will grow into trusting, easy-to-manage adult goats.)

Your goats will adapt to whichever choice you make, as long as you make the choice in advance and separate the doe and kid immediately after birth, if that is your decision.

A doe and her kids enjoying an afternoon rest.

The Goat's Preparations

As your goat enters her final stages of pregnancy, she will also prepare for the birth. Physical and behavioral changes will indicate to the goat farmer that the birth is not far away.

Humming: About two weeks before the birth, as her kids are growing inside and taking up more space, a pregnant doe's breathing will become more labored. When the barn is quiet and the pregnant doe is lying down, you will hear the change. She will breathe in normally, and on the exhale she will make a little hum, as if she is talking to her unborn kids. The more goats you have, the more humming you hear. It's very soothing to listen to.

Social interaction: Interacting with your goat may be less soothing. You may notice some behavioral changes in your goat as she nears the birthing hour. If she is normally friendly and the first to the fence in the morning, she may begin to hang back a bit and keep to herself. On the other hand, you may have a less-friendly goat that suddenly seems to like your company.

Struggling to get comfortable: Also during these last weeks you may find your goat propped up on her front knees when she lies down. She is trying to find a comfortable position as the growing kids crowd her. In fact, you may be able to see the babies kicking inside their mother. This is most noticeable on the goat's right side, near the top where her belly is bulging out. You can sometimes make out little nubs or lumps, and if the moment is right, you will see them move around suddenly.

Nesting: Often a pregnant doe will build a birthing nest as her time nears. Her natural instinct is to birth in a comforting and safe environment. All goats are different, and nest building is not a good indicator for due date, as some does will make a nest far in advance of giving birth while some will start nesting as the baby's feet are emerging. She will paw the ground and turn around and paw some more until she has a nest. Some goats will make nests all over the place, trying to decide where they feel most comfortable. Some goats just aren't homemakers and can't be bothered with nest building.

Udders filling: Udders are an even trickier sign of impending birth. With young, first-time mothers, the udder usually starts developing, but not filling up, near the end of the second month of the pregnancy. Some start producing milk two weeks or even more before birthing. Some fill up just days before birth, and still others start the actual birthing before their milk comes in. Her udder may fill out but not fill up until just after birthing. In many cases, it may swell up to such proportions that the stretching skin becomes shiny.

Rump changing: Get in the habit of looking at your goat's rump every day, examining the area just above the tail for changes. In the final stages of pregnancy, this area will be a nice round, meaty rump. When your goat is very close to birthing, however, her ligaments will loosen and you will suddenly be able to very clearly see the spine as it attaches to the tail. The muscles on both sides of the spine will drop, and you will be able to put your fingers on both sides, as if she has hollowed out. Birthing is not far away, usually within twenty-four hours.

Mucus: Just before birthing, the goat will produce a long, clear-white gloppy string from her vagina. This slippery mucus will aid in delivery. It is the final sign that the birth canal is preparing for the arrival of the babies. A few days beforehand you may see a small amount of white liquid, but not until it hangs from her backside— sometimes 6 inches (15.2 cm), sometimes all the way to the floor—do you know the kids are on the way.

Reproductive System of the Goat

and Supporting Organs

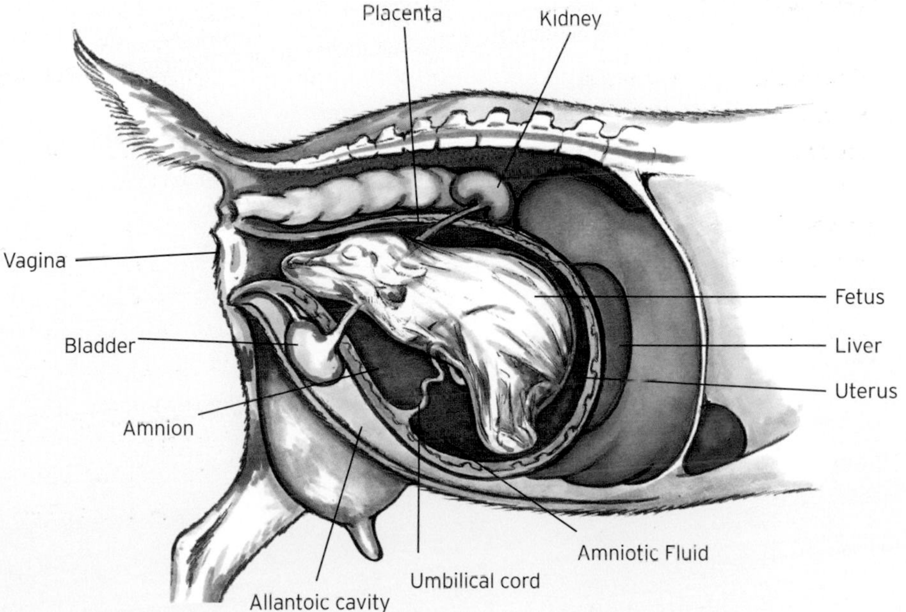

A Typical Birth

From the time you see the long white mucus, the birth is anywhere from a few minutes to several hours away. The doe's water bag will appear next, looking like a round, translucent pink balloon coming out of her. Look closely and you may be able to see the kid's feet. Usually the water breaks on its own as the kid emerges or as the doe lies down and stands up. You can break the water bag yourself by pinching the sac or poking it with a piece of straw. If it doesn't break on its own and you need to assist with the birth, you will need to break the water bag.

Your doe may give birth standing or lying or sometimes sprawled out on her side as she pushes through the contractions. If the kid is positioned correctly, the front two feet present first with head lying between the legs in a diving position. Each foot has two "toes," or portions of the hoof. If the kid is in the correct position, the toes will bend down toward the doe's feet. It's very important to note which way the toes point so you know if the kid is presented correctly.

Once feet are out, the doe will go through a series of contractions and loud bleats as she pushes. Just be patient and let nature take its course. Her body is relaxing and readying itself for passing the head and shoulders.

Closely following the feet will be the kid's nose and most likely a tongue. If your doe is taking her time and you see only two feet and start to wonder if the kid is lined up correctly, you can slide a finger inside the birth canal. You will most likely feel the kid's teeth. If you feel teeth, all is okay. The doe just needs time to relax to push the head through.

With the head out your doe may pause. She may stand up and look behind her to see where the kid is. The kid may even shake its head and open its eyes and look around. The kid is receiving oxygen from the umbilical cord still, so don't panic if you see a kid blinking at you. After the head, the shoulders will be the next difficult push for the doe. Once the shoulders pass, the kid usually just slips right out.

From first signs of feet, the birth typically takes ten to thirty minutes. Some births naturally take longer, which is not a cause for concern if the kid is correctly positioned and the birth is progressing, if slowly.

In the case of twins and triplets, the second and third kids are usually born very quickly after the first, following the same process.

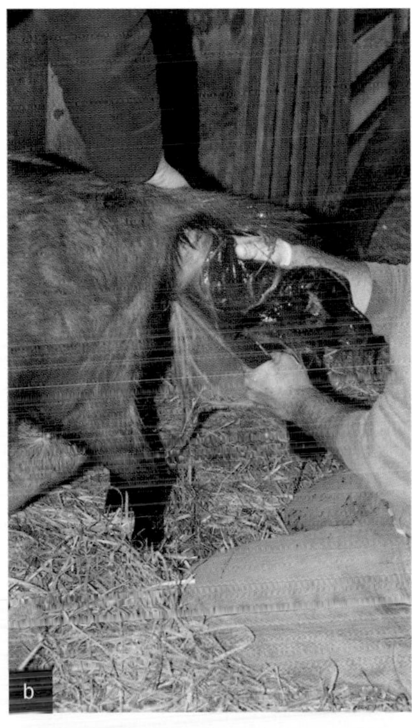

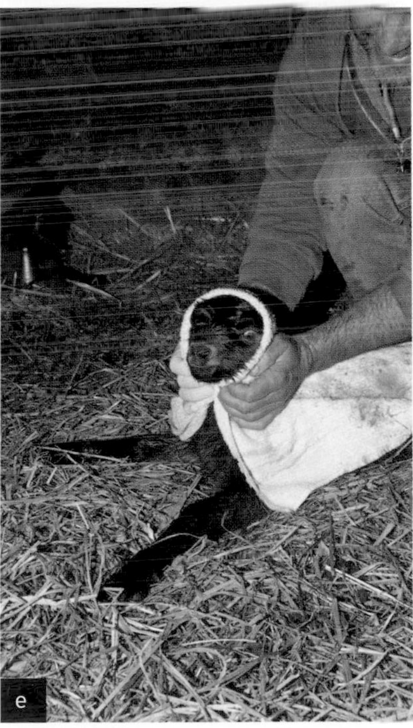

[a] This particularly large kid needs assistance getting his shoulders through the birth canal.

[b] Grasp the front legs firmly. Once the shoulders are out of the doe, reach behind and pull gently from the back.

[c] Pull downward, toward the ground. The kid should slip right out.

[d] A doe eats the afterbirth. Although we may find it repulsive, it's perfectly normal behavior for a goat. If you prefer she didn't eat it, you should remove the placenta immediately after expulsion and dispose of it properly by burying it.

[e] Here is the happy result of a successful assisted birth—a very large, 22 pound (10 kg) buckling.

Potential Complications

Most births will progress naturally, but it is important to know and recognize the most common complications: incorrect birthing position and failure to dilate.

Incorrect Birthing Positions

When a kid is not in the correct birthing position—diving out, front legs first with the nose between the legs—it can be very difficult, if not impossible, for a doe to deliver on her own. A little help from the goat farmer or veterinarian to reposition the kid or kids within the birth canal can save the doe and the babies.

The less invasive you are with a doe the better, but sometimes it will be necessary to reach into the birth canal. If it is necessary, be sure that your fingernails are clipped short, wash your hands in antibacterial soap, and use surgical latex gloves. Using one hand, tightly clasp your fingers and thumb together as if you were making a shadow art "goose head" and slowly insert your hand into the vagina. Go slowly and let your doe adjust to your hand. As you enter the doe, be aware of what you are feeling with your fingertips. In most cases, you will find a foot. Follow the foot to the leg. Follow the leg inside until you come to the breastbone, find the other leg, and find the nose. Once you have determined the position of the kid, slowly work to reposition him or her.

Here's one very important note: Never pull on the kid until you know he or she is in a suitable birthing position. What follows is a list of many potential incorrect birthing positions (what is presenting at the opening of the birth canal) and how to remedy them for a successful birth.

Position: Nose, two front feet but two different goats (Multiple kids)

Remedy: An example of why you should never pull at the first sign of feet. Although rare, it can happen that two feet are presented with a nose, but those feet aren't of the same kid. If you see two feet that don't look the same size or one foot is darting in and out while the other lies there, you may have two kids trying to exit at once. You must follow one leg in and locate its matching leg and the attached head. Determine if it is the kid closest to the opening. You may have to gently push the other kid back a bit to get the first one out.

Position: Nose, one front foot

Remedy: If you see only one foot and a nose you will have to slip a finger inside the doe and find the other foot. Sometimes the foot gets blocked in the passage and remains inside. The doe will have a very difficult time passing the kid like this, as the awkward leg position will widen the already-large shoulder area, making her strain all the more. If you feel the toes, simply grab them and extend the leg outward to normal position. If you can not feel the other foot, follow the kid's neck down to where the other leg should be, hook a finger under the kid's armpit, and slowly reposition the leg.

Position: Nose, no feet

Remedy: If a nose is presented with no feet, the feet may be pointing in toward the doe instead of out. Slip a finger in under the kid's neck and follow the body until you find the legs. The toes may be tucked under the chin. If you feel the toes, grab them and extend the leg outward to normal position. If you cannot find the toes, follow the kid's neck down to where one leg should be, hook a finger under the kid's armpit, and slowly reposition the leg. Repeat with the other side.

Position: No nose, two front feet

Remedy: This may not be a problem. With your hand, follow the legs inside the doe. If you find a set of teeth, the kid is in proper position. Your best bet is to wait until the doe contracts again, naturally moving the kid's head into position. If you don't feel teeth or a nose, the head could be bent backward. You will have to gently bring the head into place. The doe will not be able to deliver with the head bent backward.

Position: Three front feet (Multiple kids)

Remedy: Pick a foot and follow it in until you come to the breastbone, move over to the other connected foot, and draw your hand out to find the matching pair. Check to see that there is a nose and get that kid out first. You may have to reposition the kid with just one foot emerging in order to get the first one out.

Position: Two noses (Multiple kids)

Remedy: Rare, but it does happen. Again you will have to reposition the kids. Pick a kid and follow the neck down to find where the legs are. Make sure the legs are in correct position and both legs belong to the same kid. Decide which kid is first in line and if necessary push back the second kid in order to get the first kid out.

Position: No nose, no feet, just fur

Remedy: This is a breech birth. It is usually possible to reposition for normal delivery, but a novice farmer may want to rely on a veterinarian. You will have to ease your hand inside, determine how the kid is positioned, and try to reposition. In this case, it's often a kid's rear with its legs tucked under that presents first. You will need to get the legs straightened out before birth is possible. Keep in mind that with twins or triplets there will not be a lot of room for repositioning, so stay calm and work methodically and slowly get the kid or kids out. When in doubt, call your vet.

Position: Two back feet

Remedy: This is a dangerous situation; one that must be remedied immediately. Due to the positioning, the umbilical cord may be stretched or twisted, cutting off oxygen to the kid. If you see back feet (toes pointed upward toward the doe's tail), get that kid out of there. Simply grab the two back feet and gently but firmly pull.

Failure to Dilate

The cervix, the opening to the uterus, must dilate for a proper delivery. If the cervix doesn't dilate, it is impossible to get the kid out without causing extreme pain and damage to the doe. A warning sign of this condition is two feet moving in and out of the birth canal without any sign of a head. If you suspect that your doe is not dilating, don't delay. Run a finger inside the birth canal along the kid's leg in search of the head. If the cervix has not dilated, you will find a thin layer of membrane stretched tight around the legs instead. Call your vet immediately. Your vet may be able to administer medication to help with dilation. In extreme cases, a Cesarean section will be required.

Immediately after delivery, new mothers will lick their kids clean to stimulate blood flow and warm their babies as they adjust to life outside of the womb.

Postdelivery Care

Congratulations to you and your doe. You have a kid!

The Kid

You have already made the important decision to keep the kid with the mother or to separate them. Either way the kid will require your attention in the hours after birth.

Immediately after a new kid is born, depending on your climate, it may be necessary to warm him or her with a hair dryer, being careful not to burn the baby or aim the hair dryer near the baby's face, which can make breathing difficult. A hair dryer will not dry the kid as it will your hair. The kid will be covered in birthing fluid and will also need to be towel rubbed, which will also stimulate blood flow and warm him or her up.

Tie dental floss around the kid's umbilical cord 1 inch (2.5 cm) from his or her body and cut the umbilical cord. The umbilical cord that remains attached to the baby should then be soaked in iodine to prevent infection. The kid's feet can also be dipped in iodine so there is no risk of absorption of bacteria through the feet. You'll notice that kids are born with a gelatinous covering on their hooves. This protects the mother during birth and will usually slip off shortly after delivery.

Together

When you return the cleaned kid to his or her mother, the doe will naturally start licking him or her to stimulate blood flow, and most likely she will be grunting to her baby. If you have more than one goat in the pen, the other does may start licking the kid as well. Be watchful that the mother isn't pushed away by a more aggressive doe intent on claiming the baby.

The kid will attempt to stand up within just a few minutes of being born, and as soon as he or she masters standing on all fours, he will go instinctively to the teat. It's important that the kid drink the first milk, the colostrum, as soon as he is able, typically within the first twenty minutes of life. The first milk will provide the kid with the particular antibodies that he or she will need for getting a good start on life in your barn. The mother only makes colostrum for a short time after giving birth with the most potent being in the first few hours. Her colostrum-producing ability and the kid's ability to utilize it quickly diminish, so make sure your kid gets its colostrum as soon as possible and several times through the first day. If you are hand-milking your doe, you will see the colostrum, thick and yellow. When the milk turns white and frothy, the colostrum is finished.

You probably don't need to intervene in this process. After a few minutes of getting accustomed to life on the outside, the kid's hunger and instinct will take over. Better a kid that waits a few minutes and then suckles with lots of energy than a kid who eats too soon with very little interest. Once the kid shows an interest in the teat, you may have to help him or her find the nipple if you have a doe with a particularly full or low-hanging udder. If you squirt a tiny bit of milk on the kid's lips, he or she will most likely start suckling. If the teats are just so big there is no chance the kid can eat, you will have to milk a small amount into a bottle and feed the kid yourself.

Apart

If you are removing the kids at birth, you should take the kids before the doe has a chance to lick, nurse, and get too attached to the baby.

When you remove the kid and place him or her in the prepared kid box, the doe will likely sniff the birthing spot repeatedly. Some does are calm and will call for their babies for a few minutes, then move on to the hay feeder. Others suffer separation a bit longer and will call for their kids for two or three days or longer. Does that are separated from their kids may "adopt" you as their mothering instincts kick in, following you around closely. When you are milking the doe to feed the kid, she will be interested in licking your face and ears and she may talk to you, as she would her kid.

For the separated kid, you become the mother. The kid box you prepared must keep the kid warm and dry, and you must bottle feed the kid with his or her mother's milk at regular intervals to keep the baby fed but not full throughout the day. Expect to bottle feed about six times a day for the first two days, four times a day for the next two weeks, and three times a day from about week two to week eight. After eight weeks of age, you can cut the milk down to twice a day as long as he or she is eating hay and feed and consuming water from a bucket.

The Mother

Once the baby is safe and dry and warm and active, turn your attention to the mother. Have a clean bucket of warm water for her to drink. You can add a small amount of molasses to reward her and encourage her to drink.

Tip

Bottle-Feeding Schedule

Just like a human baby, a kid must be bottle fed several times a day. The average kid should be fed on the following schedule. Be attentive to your kid to ensure he or she is eating enough solid food before weaning.

Birth to two days	Bottle feed six times a day
Two days to two weeks	Bottle feed four times a day; introduce hay at one week
Two weeks to eight weeks	Bottle feed three times a day; introduce pellet feed at two weeks
Eight weeks to ten weeks	Bottle feed twice a day; free choice: hay/pellet/water
Ten weeks	Bottle feed once daily
Twelve weeks	Stop bottle feeding

Afterbirth: Soon after birthing, the doe will pass the afterbirth, the placenta. Some does pass it immediately; some take up to a few hours. It's very important that she passes the afterbirth, so keep a watchful eye out for it. Retained placenta is very dangerous to the health of your goat, as the rotting afterbirth will decay inside her, releasing dangerous toxins. When it does pass the placenta, the doe most likely will eat it, a natural instinct that is acceptable, although not pleasant to watch. You can also remove and dispose of it, most often by burying it well so that it doesn't attract neighborhood dogs, which can get sick from it.

Bouncing: If your doe has had a kid or two and now seems back to normal but still hasn't passed her placenta, you should perform an examination known as "bouncing" to determine if she has any unborn kids. Standing behind your standing goat, reach around both sides under her belly as far as your arms allow. Lift and lower her belly gently three or four times. If you feel a soft, pliable belly, chances are she has finished birthing. If her belly is taught and lumpy, she probably has another kid, so be patient and she will probably start birthing again soon. If the placenta passes, there is no need to do the bouncing test.

The udder: In many cases, the kid finds the mother's teat easily, and milk flows. If the mother suffers from hard udders, a sign of some congestion in the milk flow, you may need to milk her to feed the baby by bottle and then massage the doe to stimulate milk flow.

To massage the udder, start with two buckets of water, one cold and one warm, with a towel soaking in each. Starting with the cold towel, wrap the udder completely and gently massage the mass with lifting, probing, circling, soothing movements. Next take the hot towel and do the same thing. Alternate between warm and cold to stimulate blood and milk flow. After a couple of rounds, milk out any milk that may have gone to the nipple. Save all the milk for the kids.

CHAPTER CHECKLIST

- ☐ Have you written your vet's phone numbers in big numbers near the goat pen?
- ☐ Have you decided if you will keep doe and kid together or apart?
- ☐ Have you collected everything for your baby kit?
- ☐ Have your read and reread the information about birthing positions so you are prepared to recognize and assist with an abnormal birth?

This doe who has delivered two large kids will continue cleaning and nuzzling them until they are on their feet searching for the teat.

From the Farm: Christening P. U.

Now imagine fifty pregnant does and one farmer: Me, my first year of full-time goat farming.

The kids were all due in late January, and they arrived on schedule, fast and furious at all hours of the day and night. As I went about my midwife duties with one doe, I watched all the other does carefully for signs that they would give birth next. One particular pregnant doe, to my surprise, was watching me right back. She stared at me. I stared at her.

As you gain experience with goats, you can predict when your does will kid just by watching behavioral changes. Exhausted after a long day of birthing kids, I gave the staring goat another once-over at midnight. She was happily chomping hay and wasn't showing other signs that her kid would come soon, so I went up to the house for a few hours of much-needed sleep. Four short hours later I returned to the barn.

The staring doe was fully outstretched on the floor with very large kid lying still behind her. The doe had popped a hip during the strenuous birth; the kid had died in the cold. Between the doe and the kid, there was a big, dark-red sac. It was the doe's uterus. I had never seen a prolapsed uterus before.

Instinct took over. I rushed to the milk house for a blanket and a hair dryer to warm the doe. I left frantic messages for my sleeping veterinarian. Then I washed the uterus and tried to gently push it back into the doe. Each time I tried she would have a natural contraction, pushing it right back out again. But I wasn't willing to give up on this strange little doe.

At 6:30 a.m., the veterinarian arrived. He took one look at her and told me, "No, Brent, she has suffered too much. You have to put this poor thing down." I think the look in my eye convinced him that we need to try harder. We eventually succeeded. The vet stitched her up, prescribed antibiotics, and said, "Even if she survives, she won't be able to have kids again. Ciao, Brent!"

For two weeks, the doe lay there. I fed her by drip, milked her on her side, even helped her pee. I had almost given up when she started moving her legs a little bit, nibbling at solid food, and drinking on her own. When she finally rejoined the milking herd, I wrote "P. U."—prolapsed uterus —on her collar so I could keep a special eye on her.

The next year, P. U. gave birth to twin buck kids. Since then, she has given the farm nine more buck kids, two doe kids, and more than two gallons of milk every day. Resilient P. U. is the best goat in my barn—and she still stares at me.

CHAPTER 10

Milking

Grab your buckets—it's milking time!

Goats, like all mammals, humans included, produce milk to feed their young. With our goats, whatever milk is not consumed by the kids is ours for the taking.

If you are hoping to drink your goats' milk or make cheese, butter, ice cream, fudge, or soap, first you have to milk your goat. Milking is easy once you know how to do it. And it can be fun. You may not always look forward to twice-daily milking, but once you sit down with the goat, the process can be a soothing one, a chance to commune with your animals.

In addition to using the proper milking technique, it is important to understand how to keep the milk clean, fresh-tasting, and safe for consumption.

Healthy Goats, Healthy Milk

We want milk only from healthy goats. Milk from an unhealthy goat may not be safe for consumption. Use good judgment. If you have any suspicions about your goat's health, do not use the milk for drinking or for making cheese or other consumables until you have the milk tested. These tests are for our safety and our goats' health. Our priority is keeping our goats in good health. If our animals are otherwise healthy but a bacterial count results in milk that is unsafe for consumption, the milk can still be used for making goat's-milk soap.

A stainless-steel or high-grade plastic milking pail is a good investment for durability and cleanliness and should be reserved only for milking.

Goats love being outdoors browsing and eating the new-growth spring grass.

Three tests are available to confirm that the milk is safe for drinking.

Brucellosis test: Brucellosis is an infectious disease caused by bacteria and passed from animal to animal. Humans can be infected when consuming products from infected animals. Areas with higher risk of infection are the less-developed areas of the world including Latin America, the lower Mediterranean, the Indian subcontinent, and parts of the Middle East, but brucellosis can and does happen anywhere. Blood testing will show if your goat is a carrier of brucellosis. (See chapter 11 for more information on brucellosis.) Do not buy or keep animals that test positive for brucellosis.

Bacterial load test: This milk test evaluates the bacterial count and lets you know if you are higher or lower than acceptable bacterial loads recommended for your region. Contamination usually occurs through poor sanitation and improper milk handling rather than a sick animal. Milk will never be bacteria free. Check with your local health department to learn what level of bacteria is still considered safe for human consumption. High bacterial counts are almost always the result of human error. If your bacterial counts are high, improve your milking technique and sanitation practices and use faster cooling methods and a cleaner milking environment.

Somatic cell count test: This milk test is a good indicator of mastitis, an inflammation of the udder. Mastitis is an infection in the udder that should be treated immediately. If you want to make cheese, you will need milk from mastitis-free udders. Cheese won't set if the udder was infected, and we wouldn't want to eat cheese made with milk that has been mastitis-infected.

How to Test Your Goats' Milk for Safety

To test your goats' milk, first buy sample containers from a pharmacy. These will be individually wrapped little plastic jars with sealable lids, usually with a label. Determine if you will be testing an individual goat's milk or the combined milk your goats are producing. If you are drinking milk only from one goat, there is usually no reason to test the goats who are not milking. If you are milking two or more, you can do a group test (milk combined from all the goats).

When collecting milk for testing, make sure the teat and udder are extremely clean and your hands are washed. If you are testing an individual animal, milk the first few squirts out and throw away that milk before milking directly into your test jar. If you are testing the combined milk from your backyard farm, collect a sample from your holding tank (a jar or commercial milk refrigerator). Test the milk at the stage you would normally use it. For example, if you use the milk daily, test fresh milk. If you normally wait three days before using the milk, test a sample that is three days old. Make sure any utensil you use for transferring the milk to the test jar has been sterilized and rinsed thoroughly.

Milk can be tested at an independent lab or local health department. Testing typically takes several days. If you suspect something is wrong, milk should not be used until test results are in. If test results come back negative, continue as you were! If results should come back positive for somatic cell or bacterial load, ask your lab or vet what you can do to improve and after making changes, test again.

Skin pigmentation gives this udder an odd look, but we are more concerned about what's inside! There are tests that can be done to make sure your milk is safe for consumption.

Where to Milk

Set up your milking area apart from where your goats sleep. The cleaner and more dust free your milking area, the better. Milk straight from the goat can grow bad bacteria, so cleanliness is rule number one. A room away from the barn is ideal (and is obligatory if you plan to sell your milk and cheese as a licensed dairy). Lacking that, your milking area must be away from the goat lounging area and as dust free as possible. Outdoors is fine if the area is shielded from wind and sun. Otherwise, construct a milk stand in your goat shed.

A milk stand is a raised platform that the goat walks up or jumps up onto. Between 12 and 18 inches (30.5 to 45.7 cm) off the floor, the milk stand improves cleanliness and saves the milker from bending down to access the udders. A head restraint for the goat is usually built into the stand with a place for her food. Sticking her head through the restraint to get her food "locks" her in place until the milking is done. The milker can sit on a stool beside the stand or depending on how the stand is made, it may have a bench built in. Many backyard goat keepers find a milk stand convenient for the twice-a-day milking, but it is not a necessity if you have another clean area to milk.

The milking area should have the following:

- A clean pail with a cover
- A roll of paper towels for drying off teats
- Sterilized glass jars with lids, such as those used for canning food, for storing milk
- A fine mesh filter for straining milk
- A bucket or sink to be filled with cold water

Your milk pail should be stainless steel (without seams) or food-grade plastic and should be used only for milking. The pail should be washed with cool water (to remove remaining milk without curdling it) and then hot water (to kill bacteria) before and after every use. Tilt it upside down at an angle to air-dry. Drying with a cloth will spread bacteria.

Milk pails for goats—low, wide buckets with a cover to prevent contamination—can be purchased from a farm supply store. Look at your goat and her full udders and make sure any milk pail you get will fit under her. Filters suitable for straining milk can be found in most cooking stores. The finer the mesh, the less likely any impurities, such as hair, dirt, and hay, will end up in your jar.

You need several glass jars for milk storage. When estimating the number of jars, consider how many goats you have, how much milk they each produce (use an average of 3 quarts [2.8 L] per day), and how long you will store the milk before using. To sterilize glass jars and filters, wash them in hot water, allow them to air-dry, and dip them in an unscented chlorine bleach diluted with water (about 1 teaspoon bleach per gallon [3.8 L] of water) and air-dry.

Goat Keepers' Glossary

Goat show: This is a gathering of goat keepers and goat fans for a little friendly competition over whose goat best represents the breed standard. Not all goat shows require registered goats. Find out from your local 4-H or goat association what the requirements are for showing your goat.

This goat is alert and curious to what is happening on the other side of the fence.

Milking Routine

Once you have set up your milking area, you are ready to start your milking routine. A routine is good for you and for the goat, especially when you—and perhaps the goat, too—are a novice milker. It will help you both to relax. Relaxing helps the goat let down her milk, allowing for easier, faster milking. Consistency is key. While you are both learning, milk at the same time in the same place every day until she learns her routine. Offer her the same food—this is a perfect time to feed her morning and evening ration of pellets—to keep her occupied while you milk her.

Preparation

After the goat has come to the milking area and is happily eating her food, follow these steps to prepare her for milking.

Step 1: Quickly rub all over the udder with your hands. This loosens any dust, hair, or other debris stuck to the udder. If it is not removed, it could fall into the milk bucket.

Step 2: Massage the udder.
If you have ever seen a kid, lamb or calf suckle from its mother, you may have noticed it shoving its nose quickly into the udder. It looks like it hurts the mother, but it stimulates her to let down more milk. It is the baby's way of letting mom know he is still hungry. By mimicking that motion, we naturally let our goat know we want milk, starting the letting-down process. Ball up your hand into a fist and place it under the udder. Press your fist gently upward into the udder on both sides of the goat three or four times. This should be pleasurable for your goat and encourage milk flow.

Assess the pros and cons of purchasing an automated milking machine. It will change your milking routine. You must be certain that the advantages (financial, space, time invested) outweigh any disadvantages.

Step 3: Clean the udder.
With a wet cloth or paper towel, wipe down the udder, concentrating on the teats. Use the cloth or paper towel only once. Sharing the cloth or paper towel among goats could unintentionally spread disease from one to another.

Another Option: Machine Milking

Automatic milking machines take most of the physical labor out of milking your goats. The machine consists of a plastic cup that fits over the goat's teat with a rubber line connected to a vacuum pump. The pump vacuums and releases in quick succession, simulating the suckling of a kid goat. The doe releases the milk, and it is pumped through the rubber line into a sealed container that can then be detached and taken into the house or creamery. Milking machines come in a variety of sizes capable of milking one or two goats at a time all the way up to elaborate milking parlor systems able to milk twenty goats or more simultaneously.

Depending on your physical capabilities, milking only one or two goats does not merit the cost of an automatic milker. They are not cheap nor are the replacement parts. If you have a large number of animals to milk twice a day, the price may not seem so high!

Advantages to buying a milking machine:

- If you are physically limited in your milking abilities, a milking machine will do the work for you.
- If you have a large number of milking does, the cost may be less than if you had to hand milk, considering the time you would save.
- Milking machines, if used properly, are less damaging to the teat than improper hand milking.
- Milking machines extract the milk in a much more sanitized manner than hand milking.

Disadvantages to buying a milking machine:

- Cost—These machines are not cheap. If you are only milking one or two goats, it will be hard to justify the cost.
- Milking machines must be cleaned after every use. Your cleaning time may take longer than the actual milking time if you have only a few animals.
- If you are looking forward to that rhythmical connection to your animal and to your food, hand milking is better than a noisy vacuum pump.
- Storage—Given the cost of the milking machines, it should have proper storing area away from dust, animals, and possible breakage from accidents.

Milking Technique

The first rule in milking: don't pull! Pulling on the teat does nothing to stimulate milk flow, and you could damage the udder's ligaments or the delicate milking system inside the udder. Plus, it probably hurts the goat.

To begin milking, position yourself either behind the goat, reaching between her back legs, or next to her (either side), facing her back end. Sitting on a small stool or squatting works best. You can also have your goat hop up on a milk "stand," which will be less back-breaking for you. A stand is any raised platform about 18 inches (45.7 cm) off the ground with enough room for the goat and her food. Once the goat hops up there for her food, you can milk at a more comfortable level. Once you are properly positioned, the following steps will yield your first squirt of milk.

Step 1: Make a U with the thumb and forefinger of each hand and run a hand up each teat until you hit the juncture with the udder. This is your starting point. You don't squeeze the udder, just the teat. Hold the teat pointing down toward the ground.

Step 2: Close your thumb and forefinger together tightly to trap milk in the lower part of the teat.

Step 3: Close your middle finger around the teat, forcing the milk still farther to the lower end of the teat.

Step 4: Close your ring finger around the teat.

Step 5: Close your pinkie finger around the teat, forcing the milk out.

Step 6: Repeat.

When learning, it's easiest to alternate hands as you milk. Squeeze one teat, then concentrate on the other, while the just-milked teat fills up again. Continue to alternate and get a nice rhythm going.

Depending on your experience, how much milk your goat gives, and how big the openings to her teats are—some goats naturally gush milk out while others give tiny streams—milking could take anywhere from one minute to ten minutes. The faster you can get all the milk out of her, the better for both of you. (And you will milk only one goat at a time, straining, cooling, and storing her milk before moving to the second milking.)

Refining the Process

With a little practice, you will be able to speed up your milking and establish a steady rhythm.

Your hands and arms will probably tire quickly, but as with any workout, with practice your muscles will adjust and you'll be milking like a pro in no time. You may find that you are averaging about three minutes per goat. It sounds fast, but three minutes of good rhythm is a lot of milk and a good workout.

When milking for consumption (or for testing the milk for consumption), you should discard the first several squirts of milk. These first squirts stimulate the goat's milk and clean out any impurities in the teats. When your goat leaves the milk stand, the sphincter muscles on her teats are very relaxed, leaving them open to dust and other contaminants. After you discard this first milk, place the milking bucket under the teats, keeping the bucket covered until you begin milking again to keep it clean.

- If at any point during milking your goat becomes restless, dancing around and stepping in and out of your bucket, you need to reestablish a calm environment. Firmly but gently cup her udder with your hand and talk to her in a soothing voice until she stops moving and you can start milking again.

- While reestablishing calm, be considerate: she most likely is dancing around because she is not comfortable with what is going on back there. Be sure you are not pinching or pulling. Try to speed it up so she is not kept there for long periods.

- If she doesn't want to be touched at all, she may have a sore or an irritation. Palpate her udder to check for lumps or fever (indicating udder infection), look at the end of the teat for swelling or redness (indicating irritation or infection), and check for cuts or splinters in the teat that may be causing her pain.

When her udders seem empty and no milk is falling into the teat, remove your bucket from under the goat, cup your hand under the udder, and lift it gently but quickly several times. You may hear the goat's stomach rumble as she lets down more milk. The teats will fill up quickly again. Milk that out. When the udder feels empty and teats aren't filling up anymore, you are done milking for this session.

Tip

Milking Hints

Keep these hints nearby during your first weeks of milking.

- Keep a routine. A routine will help the goat's milk flow.
- Don't pull at the teat and never squeeze the udder.
- It's all about the hands. If you feel your biceps bulging and your shoulders hunching up, you're trying too hard. Relax.
- Establish a rhythm.
- Relax.

Storing the Milk

Once your milking is done, it's important to clean and chill the milk as quickly as possible. The longer milk sits in a pail with a few strands of hair or dust from the barn (almost unavoidable when hand milking) at warm temperatures, the faster bad bacteria will multiply. Bacteria makes the milk taste bad and can make it unsafe to drink.

After milking each doe, you should immediately strain the milk through a fine mesh strainer into a sterilized glass jar. Close the jar and place it into a sink or bucket filled with cold water to start the cooling process. The faster you cool the milk, the less opportunity bad bacteria will have to multiply, keeping the milk safe and sweeter.

Occasionally give your jars of milk a shake or twist for quicker cooling. Ideally you should have the milk chilled to under 40°F (4°C) in fifteen minutes. Never add warm milk to milk that has already been chilled, as the rapid change in temperature can cause the milk to go bad. Once the milk is chilled, it should be refrigerated.

When your first goat's milk is strained, jarred, and cooling, start milking your second doe.

Making Your Own Cheese

Let's make some cheese! Raising your own animals, milking them, and then making cheese to feed your family and friends—this is as good as it gets in goat keeping.

Cleanliness is the key for successful cheese making. Most problems in cheese making are caused by unclean equipment. All utensils used to store fresh milk or make cheese should be sterilized in boiling water with a diluted unscented bleach and water solution prior to coming into contact with the milk. Always wash your hands before starting.

For this cheese recipe, we will offer ingredients based on 5 quarts (approximately 10 pounds [4.7 L]) of milk, yielding approximately 2 pounds (0.9 kg) of cheese. For smaller or larger batches, you can calculate the recipe ratios accordingly.

You will need:

- **Unscented bleach**
- **5 quarts (4.7 L) clean, fresh milk from your goat, properly cooled and stored**
- **Stainless steel pot with cover (at least 6 quarts [5.7 L])**
- **Cooking thermometer (Avoid using the same thermometer for checking meat temperatures, sugary, or greasy substances.)**
- **1 package mesophilic cheese cultures (available at cheese supply stores)—Cheese cultures are normally sold in tiny packets about the size of a credit card. Depending on the manufacturer, the packets may be a single dose for small home cheese making or packets that will set up to 132 gallons (500 L) of milk. Always read the dosage instructions as to what size packet to use for the amount of milk you are using.**
- **1 drop liquid rennet (available at the pharmacy, grocery store, or cheese supply store)—For this recipe you will only be using a drop or two of rennet, so one small bottle will make a lot of cheese.**
- **Cheesecloth**
- **0.2 ounces noniodized salt (When estimating salt needs, calculate 1 ounce of salt per 10 pounds [4.5 kg] of cheese. For this recipe 2 pounds (0.9 kg) of cheese needs 0.2 of an ounce of salt.)**

To prepare the equipment:

Use a small splash of bleach in a tub of water—about 1 teaspoon per gallon [3.8 L] of water—to dip all utensils that will touch the milk and cheese. Be careful not to use too much bleach. You should not be able to smell chlorine on the utensils. Rinse in boiled water and place utensils upside down on a rack to air-dry.

To prepare the milk:

Use the freshest milk possible. As noted above, 5 quarts (4.7 L) of milk will give you approximately 1 quart (0.9 L) of cheese.

You can make cheese with milk directly from your goats or you can heat treat the milk, heating it in the stainless steel pot up to 167°F (75°C) and cooling it rapidly to 70°F (21°C) by setting your pan in ice water and continuously stirring the milk. Heat treating will kill off some bad bacteria, creating a healthy environment for your good bacteria to multiply (after you add your cultures) before bad bacteria can take over. If not heat treating, you can continue to make the cheese as follows, using the milk at room temperature. (Cheese does not set consistently at temperatures below 63°F (17.2°C).

To make the cheese:
To the milk in the stainless steel pot, add a very tiny amount of culture, following the instructions on the package. Add 1 drop liquid rennet, stir well for 3 minutes, and cover the pot. Let set 18 to 24 hours and open the lid. The milk will have set into cheese with a yogurt-like consistency. There should be a small amount of water sitting on top of the cheese.

Carefully ladle the cheese into a colander that has been lined with cheesecloth, leaving enough cloth hanging over the sides of the colander so that the four corners can be tied together. Tie the four corners of the cheesecloth around the cheese and hang the cheese over a pan to catch the dripping water (the whey). Drain the cheese until most of the moisture has dripped off (about 12 to 18 hours, depending on the air temperature).

Empty the cheese into a large bowl, mix thoroughly with your hands to break up any lumps, and salt to taste.

To store the cheese:
Cover with plastic wrap, pressing the plastic wrap tightly against the cheese, and refrigerate. The cheese will stay fresh for about a week. The cheese can be wrapped in a ball in the plastic or transferred to an airtight container of appropriate size.

To eat the cheese: Blend with chives for a classic dip, add olive oil and black pepper for spreading on bread, or shape into small patties, dip in bread crumbs, and lightly fry for adding to a salad.

Bonus: Aging the cheese:
You can also shape this fresh cheese into a log and age it. Store the log at about 54°F (about 12°C) for two weeks. A humid environment will help your cheese age slower, which is better. If you are aging your cheese in a refrigerator, place a bowl of water in the refrigerator, too.

As it ages, the cheese will develop a bloomy rind (a dry, folded skin) and take on a chalkier consistency. Most likely the cheese will develop a faint white mold. Molds are natural on cheeses and actually help develop flavor. If you are concerned about molds, stick to fresh cheese!

The Milking Cycle

The milking cycle starts when your goat has her kids and continues through into the fall. Depending on the individual goat, you can expect anywhere between six and ten months of milk production. Milk production actually peaks about forty-five days after the doe gives birth. This coincides with the growing babies' needs. Near-maximum milk production can be maintained for a period of time with correct feeding and proper milking. The length of the milking cycle also depends on what you plan to do with your goat. If you rebreed her, you begin a new cycle. The resulting pregnancy will more than likely signal to her body that she should slow down and eventually stop milk production and start putting her nutrients and energies into herself and her developing kids.

Drying Off

At some point the milk production will decrease to where it's just not worth the effort to continue milking or you will decide that you want your goat to stop giving milk, which is to say, you want to stop milking. The process of slowing and stopping milk production is called drying off.

Use good judgment when drying off. If your goat is producing 3 to 4 quarts (2.8 to 3.8 L) per day, it's not fair to her—and could possibly be damaging to her—to try to dry her up too quickly. Keep milking!

If drying off is appropriate, there are two methods.

Fewer Milkings

One approach to drying off your goat is to stop either the morning or the evening milking. During the now once-daily milking don't take all the milk that she is producing. Leaving about 20 to 30 percent of the milk in the udder will signal to the goat that the milk is not needed, and she will slowly stop production. If your goat is usually giving three liters a day, stop milking when you have two liters in your bucket. Milk production will decrease, and in about a week she should be giving such a low amount of milk that you can stop milking completely.

Stop Milking

Another method of drying off is to stop milking completely. When the goat's udder fills for the next scheduled milking, milk out enough to reduce the tightness of the udder but not so much as to encourage her to fill it up again. At the same time take her off pellet food and feed her straw (clean baled straw usually used for bedding) instead of hay at one feeding each day for three or four days. She will enjoy munching on it, but it will not provide enough nutrients for milk production. Within a week, she will be dry.

CHAPTER CHECKLIST

- ☐ Have you set up a milking area that is both clean and convenient?
- ☐ Have you practiced milking? Hold one hand upside down and ball your fist, leaving one finger hanging. Practice milking your finger with your other hand!
- ☐ Have you established a routine?
- ☐ Do you understand how and why to treat the milk?

From the Farm: The Rule of LeeAnn

Through most of the 1990s and until 2004, I co-owned and ran an agriturismo, Italy's version of a farm holiday. I had big house filled with guests and I cooked breakfast every morning, with homemade bread and homemade butter from the cows.

The guests loved to experience the farm. Some enjoyed picking produce in the garden for dinner. The children always wanted to collect eggs for breakfast. Sometimes I had to put the eggs back into the nests so that the next family to wake up would have eggs to collect, too!

Occasionally, a guest would get up at 5:30 a.m. to help me milk my fifteen goats. I was happy to have the company. At the time, I had a simple lean-to shed, with a simple head-lock system built in. (This might be as simple a setup as some of you may begin with.) As the goats were fed, they were locked in place for easy milking.

You know by now that I am very attuned to my animals. I am always thinking of what the animal wants and needs, and I tried to relay my way of doing things to my guests. My helper this particular day was a lady named LeeAnn. I gently talked LeeAnn through the routine, explaining the importance of rhythm and relaxation. I explained how the goat needs to get used to the milker.

The goat she chose to milk was a beautiful Maltese we called The Flying Nun because of her enormous ears. A gentler, more ladylike goat would be hard to find. She was always patient with first-time milkers.

LeeAnn reached under the Nun and grabbed her teat a bit more directly than I had suggested, and the unhappy Nun started to prance around. I was concentrating on filling my own milk bucket when out of the corner of my eye I saw, quick as a flash, LeeAnn straight-arm slap my poor Flying Nun on the side while yelling "Stand still!"

Well, Nun was having none of that, and neither was I. I gave LeeAnn a talking to: My barn. My rules.

From that point forward, I cautioned my helpers strongly that if they would like to help, they must treat the animals as I do, with kindness and respect and absolutely no walloping. Around the farm, we still call this the "Rule of LeeAnn." And we use it as a threat: "If you don't hold still, I'm gonna LeeAnn you in a minute!"

Of course, we never have.

CHAPTER 11

Goat Health

The first step in raising a healthy goat is starting out with a healthy goat, one that has been well cared for (see chapter 6) and was born into a healthy herd. Goats are hardy animals, and with continued good care, your healthy goat is likely to remain in good health.

Good care includes proper feeding and access to the additional vitamins and minerals a goat would naturally seek out on a pasture. Your backyard goat will find these vitamins and minerals in her pellet food, forage, hay, limited salt block, and baking soda as well as through access to sunlight. Good care also means regular exercise to keep your goat's mind and body active and necessary maintenance, including both hoof trimming and brushing (see chapter 7) and regular visits from the veterinarian for deworming.

Beyond the basics, there are two schools of thought when it comes to goat health. Some goat farmers and veterinarians will advise you to use medication as a preventive measure, treating your goats for diseases common to the area before they become ill. On the other side of the fence are goat farmers who take a natural approach to prevention and administer medications only if the animals become ill. Neither approach has a 100 percent success rate. Talk with your veterinarian about how you are using your goats and which diseases are prevalent in your area.

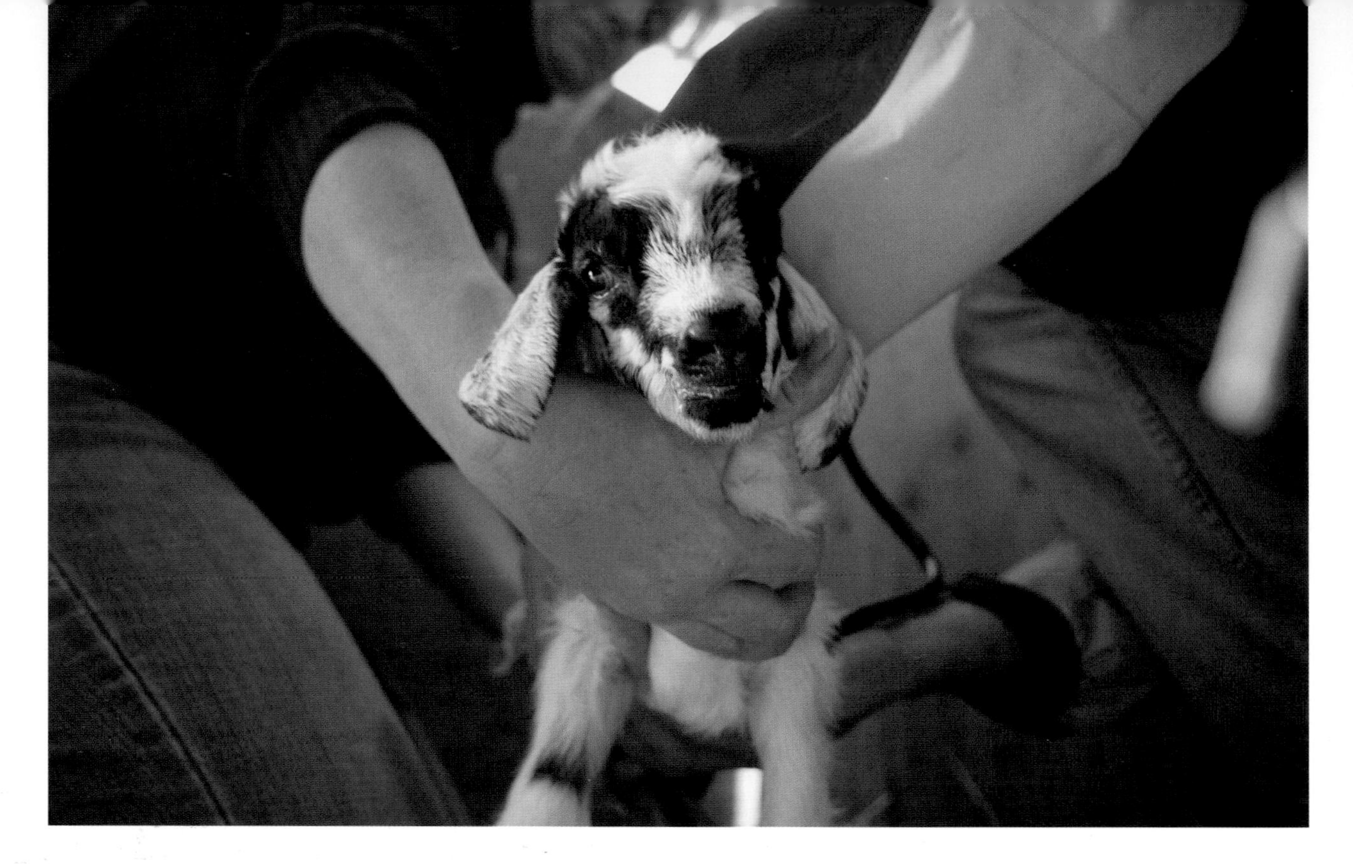

Always have your veterinarian's phone number in an easy-to-remember place, and don't delay in calling it if you have any doubts about your kids' or goats' health.

Choosing a Veterinarian

Ask the goat owner you buy your goats from if he or she can recommend a trusted veterinarian for your goats. With the growing interest in goats and goat keeping, every year the number of veterinarians who specialize in small ruminants increases. There is also a growing community of goat owners on the Internet happy and willing to share advice and experience, often offering home remedies for minor health issues or writing about their goat problems, which may be similar to yours. Your local health department will either have or give you a contact for veterinary services.

If possible, your veterinarian should be a small-ruminant specialist. Ruminants with their multiple stomachs have different dietary needs (as discussed in chapter 5), and not all vets are familiar with those. Also, if you are going to be milking your goat, a veterinarian who is familiar with the mammary system of milking animals will be valuable.

Keeping a Health Diary

There is the average goat, with a heartbeat of seventy to eighty beats per minute and a body temperature of 102°F to 103°F (38.9°C to 39.4°C). And then there are your goats, which might have a slightly different normal heartbeat or temperature.

A health diary will establish a picture of your goats when they are healthy. Goats are all different, so be sure you don't confuse your goats in the health diary. It's easy to set up an organized system when you have only a few goats. Start with a new notebook. On the front inside flap tape a copy of the average goat vital signs as noted in this chapter along with the number of your veterinarian. Keep a separate section for each goat starting with her name, date of birth, and any other information you might feel is important about her (for example, she was a twin, a runt, a big baby, etc.). Underneath record her temperature readings to establish her "normal" temperature. Leave several pages blank that you will use in the future to update any veterinary visits, blood or milk tests, pregnancy complications, births, and other observations.

Monitoring temperature: Take your goat's temperature throughout the day and throughout the seasons and record the results in the health diary. A goat's temperature may vary due to factors such as weather, housing, season, and movement. Digital rectal thermometers that beep when ready take the guesswork out of getting an accurate reading. If your goat's temperature is outside her normal range, you should call your veterinarian.

Tracking medications: You should also use the health diary to record any medications you or your veterinarian administer to the goat. Note which medications were administered, how, when, to which goat, and why. You can also use the health diary to store medicine labels. This is especially important if you are using the animals for milk or meat. The labels will tell you how long the medicine will stay in the goat's system and when the meat and milk will be safe to consume. For instance, the label may say "Waiting period for milk: 72 hours."

Goats need a hay feeder of the proper height with enough space surrounding it to allow more than one goat to feed at a time.

Before You Buy a Goat

Dangerous diseases can spread among goats in a herd or pass from doe to kid. Before you buy a goat, be sure that the goat has been tested for and is free of the following diseases. The goats are tested individually, and normally if the doe and rest of the herd are clean, the kid will not need to be tested.

Brucellosis

What it is: It's an infectious disease caused by bacteria and passed from animal to animal. Although it is becoming increasingly rare, brucellosis can be transmitted from goats to humans through the consumption of raw milk. Ask your local health officials, veterinarian, or the goat seller if brucellosis is a concern in your area.

Testing: Dairies and licensed cheese-making farms will be required to test for brucellosis before selling their milk and milk products to the public. In Europe brucellosis tests are mandatory before transportation of animals between farms. Drawn blood from each animal is sent to a lab for testing.

Caprine Arthritis Encephalitis (CAE)

What it is: This is a very serious disease in which infected animals may waste away; their joints may become swollen, causing mobility and health issues. In younger animals, encephalitis may develop. This virus is passed between goats through body secretions. There is no evidence that CAE can be passed to humans, but dangerous pathogens can, so be sure to drink only from healthy goats.

Testing: Goats are individually tested with a blood sample. Goats may test positive and not show symptoms until years later, sometimes not developing symptoms at all. There are preventive measures to keep kids CAE free, immediately separating them from the doe at birth before she has a chance to clean or nurse the kids and heat treating the mother's colostrum and milk. The best prevention is to start with kids that have been raised CAE free from CAE-free does.

Johne's Disease

What it is: This is a gastrointestinal disease that thickens the wall of the intestines, making absorption of nutrients difficult and causing rapid weight loss even though the goat has a healthy appetite. Goats can be infected very early in life but not show symptoms until years later. The disease can be passed by body secretions, drinking troughs, even passing from doe to unborn kid.

Testing: A blood test will show if the goat has the Johne's antibodies, meaning it has been exposed to the virus. The best prevention is buying your goats from a Johne's disease–free herd.

In Your Barn

Sometimes, despite proper care, goats will develop a disease or condition in your barn. Be prepared to recognize and quickly address the following issues.

Abscesses: These are large, pus-filled lumps that can occur anywhere on a goat. Very much like a large pimple, they will swell with pus until they break. Some abscesses—known as caseous lymphadenitis—are contagious and dangerous to the health of your herd. Caseous lymphadenitis abscesses are usually located in the lymph glands but can also infect lungs and other internal organs, which may eventually cause death to the animal.

Prevention

Some abscesses result from getting a shot, an insect bite, or a minor wound. These are not dangerous. If you do not know the source of an abscess, separate the affected goat from the herd as a precaution.

Treatment

If you catch the abscess before it breaks, separate your goat, clean the abscess, and using rubber gloves, "pop" it. Squeeze the pus into a sample jar to give to your vet for analysis. Clean the ruptured abscess with iodine and throw away the gloves, paper towels, and anything else that came into contact with the infected fluid. Animals testing positive could contaminate your entire barn and neighboring animals. They should be culled.

Abortion: This is the spontaneous death of a fetus.

Prevention

Possible causes of early and midpregnancy abortion include worm overload, liver flukes, and poor nutrition. Late pregnancy abortions can often be attributed to your goat being rammed in the side by a stable mate or some other trauma to her belly. Ensure that your goat is in good health and isn't being bullied late in her pregnancy.

Treatment

To help prevent it from happening again, it's best to find out why the abortion occurred. You or your veterinarian can take the aborted fetus to a lab for analysis to find the cause.

Bloat: This is the sudden swelling of the rumen caused by overeating or sudden change in the diet. Your goat will swell on her left side, and the swelling will feel hard. She will look distended on that side. Gas is increasing faster than she can expel it.

Prevention

The key to avoiding bloat is careful feeding and keeping food stored far out of goats' reach to prevent overeating, should your goats get out of their confinement. Avoid sudden changes in diet. Avoid overfeeding. Free access to fresh baking soda will help your goats self-regulate their rumens.

Treatment

In light cases of bloat, you may be able to massage the gas out of the goat. If her side is still somewhat pliable, have her stand with her front end elevated and massage her sides in an upward motion. She may belch up enough gas for the trauma to pass. In extreme cases, a vet will be needed to puncture a hole in her distended belly to allow gas to escape. This is very dangerous to the goat, so try your best to avoid bloat.

Conjunctivitis (Pinkeye): This is a reddening and swelling of the eyelids causing tenderness and tearing. Dust that carries bacteria, flies, and contact with other infected animals can spread the disease.

Prevention

Separate infected animals immediately.

Treatment

Consult your veterinarian about the latest treatments. With early diagnosis and treatment, animals usually recover in a matter of days. For more advanced stages, recovery may take several weeks. Left untreated, it can cause blindness.

Coughing and Pneumonia: Coughing in goats sounds much like coughing in humans. Coughing, nasal discharge, and fever can be a symptom of damp living conditions or sudden changes in climate, such as hot, humid days followed by cool evenings.

Prevention

If your goats are coughing with some regularity, the first thing to check is their housing. Is it closed and stuffy or humid? Make the necessary adjustments to bring in fresh, clean air.

Treatment

Coughing that seems more persistent than an occasional cleaning of the lungs should be looked at by a vet immediately.

Cuts and Other Minor Wounds: Your goat may simply have cut herself on a piece of fencing or nail sticking out of a doorway you hadn't noticed or she may have been horned or attacked by another goat.

Prevention	Treatment
No farmer can prevent all accidents, but be aware of safety issues in your barn.	For less severe wounds, clean the area, apply antiseptic, and allow it to heal. For large cuts or gashes, call a veterinarian immediately to dress the wound. Do not wait, as the goat may be at risk for infection. Whatever the cause, your goat may be at risk for tetanus. When establishing a relationship with a veterinarian, ask if tetanus is a concern in your area and if possible, keep some tetanus vaccine on hand for emergencies.

Enterotoxaemia: This is when bacteria multiply in the intestine to toxic levels that are absorbed in the blood and quickly spread to vital organs. It is often associated with overeating of milk or grains. Young kids can die in very short time frames from the onset. Symptoms usually include painful groaning or crying.

Prevention	Treatment
Consult with your veterinarian for preventive medicines, which are usually given to the doe before she gives birth.	This is very difficult to cure if your kid is already lying on the ground and groaning. Prevention is the key.

Foot Rot: Foot rot is a separation and rotting of the hoof wall. It is a contagious disease passed from one goat to another through walking in the same areas. If your goat is limping and you smell something foul coming from your goat's foot, it is most likely foot rot and must be taken care of immediately before it contaminates other goats.

Prevention	Treatment
It is best to not let the bacteria enter your yard. Buy from herds that are free of foot rot.	Proper and regular foot trimming will help you detect a problem early, when it is more easily eradicated.
The bacteria that causes foot rot can live in the soil or bedding up to three weeks, so it's best to isolate your animals on clean bedding for that time period.	Veterinarian-prescribed medicines and footbaths will help clear it up.

Lameness: This is pain in the hooves. It is easy to spot. Your goat will walk strangely, favoring a leg or two.

Prevention	Treatment
The most common cause of lameness is overgrown hooves. Keep her hooves trimmed correctly and frequently. Ensure that her sleeping area remains dry and that her housing protects her from heavy rains. Ensure that her housing and outdoor areas are free from nails, glass, wires, and pieces of metal that could puncture her foot.	Same as prevention

Lice: These are external parasites that live on the skin of your goat. If you see your goat constantly scratching against the barn or a fence post or rubbing an area raw, have your vet check for lice.

Prevention	Treatment
Any animal with fur is at risk for lice. However, starting with goats from a lice-free herd makes the chances of your goats catching it slim. Plenty of sunshine, an occasional brushing, and fresh air will all help keep lice at bay.	A vet-recommended de-lousing powder will take care of the problem. You must treat all your animals at once to prevent a reinfestation.

Mastitis: This is a generic term for most problems within the udder system, including congestion of the udder, a hot udder, and lumps in the udder and either thick milk and no milk.

Prevention	Treatment
General barn cleanliness and proper milking technique are your best defense against mastitis, but it can occur despite your best preventive efforts. Consult with your veterinarian for preventive medicines, which are usually given to the doe before she gives birth.	For minor cases of congestion or hot udder, alternately massaging with hot and cold water may be all she needs. (See chapter 9.) On the other end of the spectrum, you will have your vet administer anti-infection and mastitis-fighting meds directly into the teat canal. Depending on the severity of the problem, your goat may actually lose a side of or the entire udder. Do not take the wait-and-see approach. As soon as you note changes in the udder, clumpy milk, strands of blood, or hot, hard or lumpy udders, take action with massage and place a call to your vet.

Worms: These are internal parasites.

Prevention	Treatment
Consult with your veterinarian for a deworming program. All goats are carriers of worms; the key is keeping the number of worms—the worm load—at a level that is not damaging to the goat.	Consult with your veterinarian for a deworming program. All goats are carriers of worms; the key is keeping the number of worms—the worm load—at a level that is not damaging to the goat.

With practice and lots of observation, you will be able to recognize when your goat isn't feeling 100 percent.

Should I Call the Veterinarian?

If you walk into your goat house and notice one or all of your goats looking sickly, give each a quick physical exam and try to determine the cause. If you need to call a veterinarian, the more information you can relay about their condition, the more help a vet can provide.

Simple Examination

If nothing is obviously wrong with your goat, such as a broken leg or open wound, conduct this simple exam to help find the cause.

- Take her temperature. Is it higher than normal?
- Check under her tail for signs of diarrhea. Occasional diarrhea is normal, but frequent diarrhea or a goat with a fever and diarrhea is cause for concern.
- Feel along her throat for any obstructions. Is she breathing normally?
- Check her left side for bloating.
- Watch if she brings up a cud. Cud chewing is a good thing and means she is still doing okay for the moment. Continue with the exam.
- Feel all over her body, feeling for lumps, cuts, and broken bones.
- Peel back her eyelid and check that her color is good. In healthy goats that should be red to dark pink. If her eyelids are light pink to white, she may be suffering from worm overload.
- Feel her udder. Does it feel feverish?
- Feel her feet and lower legs. Do they feel abnormally hot or cold?
- Is she reacting to you in any way or letting you examine her without any interest from her?

If you did your exam and everything came up normal, including temperature, maybe she is just feeling out of sorts and it will pass. If her temperature is high or you have discovered another abnormality, call your veterinarian.

As you gain experience, you will be better able to recognize when your goat is in trouble. Be a responsible owner and learn all you can as to what is normal and what is not. Call your goat friends, look up the many websites now available, and join a goat chat room. If you are concerned about your goat, ask for help.

CHAPTER CHECKLIST

☐ Have you selected a veterinarian?
☐ Have you started a health diary?
☐ Have you purchased a healthy animal?
☐ Are you aware of the diseases and conditions you should watch for in your barn and taken the proper preventive measures?

From the Farm

It's fun to daydream and plan your farm, what kind of goats you will have, and the products you will make with them. Milk, kids, butter, cheese—I plan it all out and even figure out how much money I can make. I could show you the "financial" spreadsheets I used to create in the early days, calculating that five goats give one gallon a day, which leads to ten pounds of cheese that I sell for . . . pure profit. Never once did I ever figure in worm loads, broken legs, bloat, or abortion.

The reality of raising animals is that sooner or later, no matter how small or large the scale, we will all have to face medical issues with them, emergency and nonemergency alike. Try as we might to give them the best care we can, there are just too many factors beyond our control. This a farming fact of life.

Being aware of what could go wrong and having a plan of action, even a simple one, like calling a vet at the first sign of trouble, is a step toward healing and possibly saving the life of your animal. Very few goat owners are veterinarians, so we can't be expected to know what to do in all situations. As you gain experience and recognize your goats' normal and abnormal behavior patterns, calls to your vet or goat advisors will be fewer and farther between. The more you read about goats, chat with fellow goat enthusiasts, exchange barn stories, and simply observe your animals, the more your confidence will grow. Little by little, you will be more prepared for each situation that comes your way.

Sometimes it feels like I have been at this forever and I can handle anything. Then there are days when I walk into the barn, notice a goat in peril, and think, "I don't know what to do." That feeling passes quickly (but it does happen!), and I move on to do what needs to be done. And that's normal: we do what we can.

No one person knows it all. Not my vet, not your vet, not someone who's been raising goats for twenty years—so don't waste time feeling overwhelmed by what could go wrong. Focus on what you need to do to keep your goat-farming projects going in the right direction and you'll be fine.

CHAPTER 12

Conclusion

You've made it this far. Goat shopping may be in your not-too-distant future. The seeds were probably planted long before you picked up this book, and if you've read carefully, you undoubtedly understand what goats are, how they function, and what they need. Now the question is, are you cut out to be a goat farmer?

Throughout this book, you've considered the logistics of backyard goat farming. You've asked yourself, can I provide for my goats in such a way that they will be a welcome addition to the neighborhood? Will my little herd have enough space and food, shelter, and shade that they want for nothing all day while I am at work? Have I thought of everything—attractive and functional fencing, a system of waste removal that doesn't offend the entire block? (The latter is necessary for preventing phone calls from the neighbors that start with, "Hi, yeah, it's about your goats ...")

Now ask yourself, will I enjoy doing all of that?

The author's goats enjoy a morning walk with views of the village and mountains beyond.

Qualities of a Backyard Goat Farmer

Every backyard goat farmer needs some things that you can't buy at the farming supply store:

Patience: You will need oh-so-much patience, not only for your goats but for the town rules and regulations you must live by and neighbors you must appease.
Humor: When dealing with goats, there will be many times when you can either curse or laugh. Choose laughter.
Dedication: Bringing goats into the family is not something you can take lightly. If you aren't 100 percent dedicated to your goat project, trouble surely lies ahead.
Fortitude: If you choose to breed your goats, be ready for late nights of kidding, the possibility of putting your hand inside a goat in case of emergency, and the need to sell kids you don't have the room to raise.
Commitment: You must make a commitment to keeping the goats in good health, commitment to being there should they need you during birthing, and of course, the commitment to the daily chores at least twice a day, every day of the year. A farmer must enjoy that commitment and participation in the seasons and cycles of life, the purposefulness of it all. Make sure you are one of those people. If you want to go on vacation, who will watch the kids? Make sure you can answer this question.
Passion: If you have passion about anything you do in life, it doesn't feel like work. Raising goats with passion makes everything so much easier, even the hard days. If the passion is there, then by default so is patience, humor, dedication, fortitude, and commitment.
Time: Being a backyard goat farmer also requires time. The time that you have to dedicate to your goat project will also determine, to some degree, the level of success your goat project will attain. You get out of goat farming what you put into it.

To be successful with goats, you need to be in it for the long run. The unfortunate economic reality for the backyard goat farmer, at least in the short term, is that your morning glass of milk will be a costly one. The cost of building housing and fencing, purchasing and feeding your goats, and all the related start-up costs are hard to recuperate in a short amount of time. The longer you have your goats, the fewer costs you will have. Keeping animals is costly and time-consuming; driving to the store for a half gallon of milk is not. The backyard farmer believes that homegrown milk is worth the expense.

Once you have goats in your yard, there will be plenty of opportunities to practice patience and commitment and indulge your sense of humor.

Goats will eat eagerly from your hand and remember you the next time you pass by!

FORMAGGI VALLE DI MEZZO
www.valledimezzo.com
Anghiari
Tel.0575-788103

Beyond Milk

If you become a backyard goat farmer, be prepared for goats to take over your life. It won't just be the chores. The chores are manageable. It will be the awesome possibilities that occupy you.

The products made from or with the cooperation of goats is impressive. The first thing we think of is milk. It's sweet, frothy, and warm—directly from the pail to the drinking glass. From milk you may venture on to make butter, healthy yogurt, cheese, and whey. We can use whey as the liquid ingredient when we bake bread, we can drink it for a refreshing summer drink, or use it to nourish our goat kids or feed other animals. Dogs and cats love it. Extra milk can also be used in making goat milk soap, a fun and profitable home business.

The kids that are born each year are either sold off for an economic return or you can learn home butchering, creating a source of healthy and home-raised meat. Then the hides can be used for tanning.

Fiber goats can provide yet another satisfying and sought-after home business. You'll find yourself not just shearing, but also spinning fiber into yarns, maybe even knitting the yarns into clothing.

It's not unimaginable that the rhythm and rewards of farming, even on a small scale, may entice you to turn every square inch of your backyard into a farm. Your free time will then be dedicated to this therapeutic and completely satisfying hobby.

A basket of the beautiful, artisanal cheeses made by the author on his farm

Building a Goat Keeping Community

You won't be alone. Goat keepers are no longer alone on a high on a hill with their herd. Online goat-farming communities connect that high hill with your backyard and other goat farms around the world. Explore goat websites and join goat chat rooms. You'll always find a friendly voice, lots of beginners just starting their adventure, and quite a few experienced farmers offering advice to help get you started in the right direction.

In your own community, introduce yourself to farmers at the farmers' market. Attend 4-H or other goat clubs to meet other goat farmers while learning about your caprine friends.

Your circle of goat-farming friends will be a great source of information. Remember: Being a farmer is about being flexible. Gather all the information you can before making a decision about your goats. All this information is a guide, and no single farmer or book, including this one, is going to have all the answers.

If you choose to join the ranks of backyard goat farmers, your goat adventure is sure to be exciting and unexpected. Have fun with it, share it, spread the word, and be a great ambassador for goats.

From the Farm: It Started with One

I wasn't always a farmer.

I went to college, and I tried to go the corporate route. But I would sit in my office, and instead of working, I would try to figure out how quickly my herd of (imaginary) goats would grow if I started with two does and kept their female babies every year. What I calculated out was that life in an office wasn't for me.

While pondering my next move, I bought a quart of goat's milk at the Greenmarket in New York's Union Square, and with that classic beginning cheese maker's guide *Cheesemaking Made Easy,* I made, unbeknownst to me, what would be the first of thousands and thousands of cheeses. It actually turned out pretty good, and I served it at a dinner party to lots of oohs and aahs.

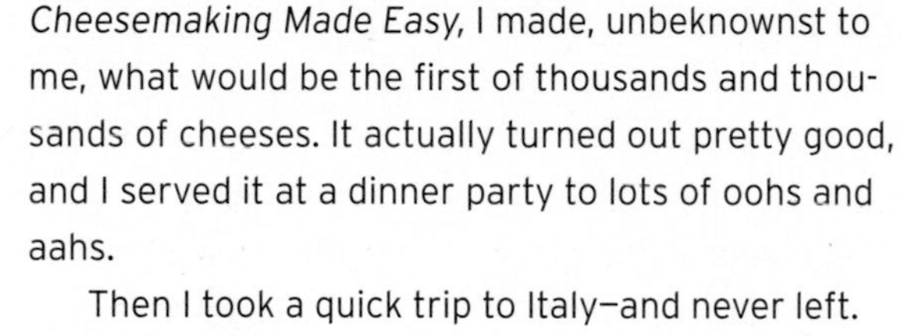

Then I took a quick trip to Italy—and never left. When my elderly neighbor ladies gave me a little Saanen doeling named Babbi, I couldn't have imagined that I would still be in Italy twenty years later with eighty goats out in the hills. My neighbor ladies, with the exception of Mamma Rosa, are still here, too.

I raised goats and cows and sheep and turned their milk into cheese over the fire in the living room of that first farmhouse. I was obsessed with cheese making. If the cheese turned out okay, I tried to sell it; if it didn't, I gave it to the pigs or the chickens or the dog.

Then came the guests. For ten years, I ran a bed and breakfast on the farm, introducing guests to the goats and the garden. The guesthouse was successful and busy, and little by little I started selling off the animals to make more time for cleaning the house and catering to guests.

With a wave goodbye, the author heads back to the barn for evening milking. Good luck on your journey!

But I missed being a farmer, so I left the guests on the mountain and moved ten miles (16 km) (down the valley to my current farm with lots of goats that give lots of milk.

I now make fourteen types of cheese, which for a two-man operation is far too many. Every year I say, "Next year I am cutting back the cheeses," and somehow, next year, I always make one or two more. Fresh, soft, spreadable cheeses, raw milk cheeses, aged table cheeses, mozzarella (at the beginning of the season when the milk stretches easily), carbon ash-covered cheese, bloomy rinds, blue cheeses, and cheese wrapped in various leaves from stinging nettles to tobacco, sold at Tuscan farmers' markets.

Of course, I can't resist sharing the farm, so the farmers' markets also give me an opportunity to invite people to visit the goats, taste some cheese, and hopefully buy some more. That has led to weekly tour groups who visit for guided tours of the barn, cheese tasting, and a chance to milk the goats. And then there are the volunteers: For the past two years, I have been hosting volunteers of all ages who come to the farm loving the idea of working with goats and making cheese. The reality of a working goat farm is a far cry from the long, lazy country days in the Tuscan hills that many people envision. Some of the volunteers fall in love with the long days. Others go back to their office jobs.

When I walked up the mountain that day with Babbi, I never imagined my life would revolve around goats. That little goat turned into several more goats, another farm, lots of cheese every week, and now this book on raising goats, which may be the start of your goat adventure.

This farmer needs to sign off now. It's getting late, and I have to be up at 5 a.m. to do the milking.

RESOURCES

All my books are rather worn from reading and rereading. I own all of the following. Every time I read them, I learn something new that helps my goat project continue to shine.

Books

Raising Milking Goats Successfully. Luttman, Gail, Charlotte, VT: Williamson Publishing, 1986. My first goat-specific book—
I can practically recite the entire thing cover to cover I have read it so many times.

Storey's Guide to Raising Dairy Goats. Belanger, Jerry, North Adams, MA: Storey Publishing 2001. This is a complete how-to guide that no goat owner should be without.

Home Cheese Making. Carroll, Ricki. North Adams, MA: Storey Publishing 2002. What cheese maker doesn't have this classic? If you plan on making cheese, this is a great starter book.

Practical Cheesemaking. Biss, Kathy. Ramsbury, Marlborough, Wiltshire, UK. The Crowood Press LTD 1988. This is an easy-to-understand technical book written by Kathy Biss, who has long taught the art of cheese making. This author knows her cheese and knows how to explain it.

The Encyclopedia of Country Living. Emery, Carla. Seattle: Sasquatch Books, 2008. Filled with how-to, good advice, recipes, and humor, this book tackles all aspects of homesteading, animal raising, gardening, and preserving. Great reading even if you don't plan on plowing with oxen or eating a bear.

Barnyard in Your Backyard, Damerow, Gail. North Adams, MA; Storey Publishing 2002. This is a starter handbook for creating your own perfect small-scale backyard farm.

The Year of the Goat, Hathaway, Margaret. Gilford, CT: The Lyons Press 2007. This is a fun and inspirational read telling of Margaret's cross country trip to learn all things goat

More Backyard and Homestead Books from Quarry

The Backyard Beekeeper
Kim Flottum
978-1-59253-607-8
Sage advice for the beginning beekeeper

The Backyard Beekeeper's Honey Handbook
Kim Flottum
978-1-59253-474-6
Everything you need to know to make the most of your bee's honey

Better Beekeeping
Kim Flottum
978-1-59253-652-8
Practical information for taking beekeeping to the next level

The Backyard Vintner
Jim Law
978-1-59253-198-1
Artisan winemaking recipes and techniques

The Chicken Whisperer's Guide to Keeping Chickens
Andy Schneider & Dr. Brigid McRea
978-1-59253-728-0
A down-to-earth tutorial on raising chickens in your own backyard

Intertwined
Lexi Boeger
978-1-59253-624-5
Spinning techniques for the aspiring fiber artist

Making Artisan Cheese
Tim Smith
978-1-59253-197-4
Illustrated step-by-step methods for making gourmet cheeses at home

Magazines

Country Folks magazine
www.countryfolks.com This magazine offers sound advice on raising and showing livestock.

Hobby Farms
www.hobbyfarms.com Here's all the essential information you need to launch and run your hobby farm.

United Caprine News
www.unitedcaprinenews.com This is a monthly magazine that offers goat-care tips, showcases breeds, lists breeders, and makes connections among the goat-keeping community.

Urban Farm
www.urbanfarmonline.com This is all about sustainable urban living from the creators of *Hobby Farms*.

Websites

www.dairygoatjournal.com This website is constantly updated with news and information for beginner or seasoned goat folk.

www.duhgoatman.tripod.com This site is arranged by state and by breed, making it easy to find breeders and raisers of goats near you.

www.fiascofarm.com You will be hard pressed to find a website on goats that gives you more information than Fiasco Farm.

www.goatconnection.com This is a great website for "everything goat."

www.goatjustoceleague.com The Goat Justice League was founded to legalize the keeping of goats within the city of Seattle. Their website offers a wealth of information for any goat lover.

www.goatkeeper.ca This is Canada's goat keeping online magazine.

www.goatsthatfaint.com
Bee Haven Acres offers sound, healthy, friendly Myotonic goats (Tennessee Fainting Goats) eligible for registration with the Myotonic Goat Registry and International Fainting Goat Association.

www.goatworld.com This website offers advice on everything from raising goats, goat breeds, housing, fencing, cooking goat meat, making soap, and classified listings.

www.hobbyfarms.com Beyond just a magazine, this site is for the small farmer in all of us—a favorite of mine with helpful information on all things agricola on a small scale.

www.seattletilth.org/learn/ See the Seattle website for information on classes and workshops or research other local options on line.

www.sheepandgoat.com See the Maryland small ruminant page with information on keeping small ruminants including feeding, disease management, meat, dairy, fiber, and fencing.

Supplies

When it comes to goats, there are things we want and things we need. These supply houses have both.

www.caprinesupply.com This is an online store catering to the goat enthusiast. It's easily organized into kidding, birthing, and milking sections offering medicines, a library, and gifts.

www.hoeggerfarmyard.com This is one-stop shopping for all things goat: books, soap and cheese making supplies, milking equipment, medicine cabinet necessities, halters, fly remedies, and more. They also offer gift certificates for the hard-to please goat fancier in your life.

Cheese

www.cheesemaking.com Here are resources by Ricki Carrol. This website is packed full of information, stories, cheese vacations, and recipes. Sign up for the "moos letter."

www.culturecheesemag.com Here are all things cheese: reviews, recipes, blogs, questions and answers, events, and cheese making news from around the globe.

Organizations

www.4h.org This is a classic organization to get children involved in agriculture, raising animals, helping communities, and learning practical skills.

www.adga.org This is the American Dairy Goat Association website with registered and recorded dairy goat performance and lineage. Information on dairy goat expos, shows, breeders, auctions, history, and promotion of the goat raising industry.

Fiber

www.joyofhandspinning.com Here's a good source of instruction, information, and supplies for the home spinner.

www.northwestweavers.org This is a regional publication of the northwest USA. It's a great source of information and contacts for anyone interested in home weaving, fiber supplies, animals. It provides sheep and alpaca information as well.

ABOUT THE AUTHOR

Brent Zimmerman has had a lifelong interest in raising and caring for animals, specializing in goats for the past twenty years. His animals-first philosophy and a true passion for what he does has landed him several guest appearances on Italy's most popular food-farming television shows. He is an active member in local farmers' markets and opens his barn to tour groups and school children promoting a local agricultural way of life and the welcome trend of "know your farmer, know your food." He lives in Tuscany, Italy, where he raises goats and makes artisanal farmstead cheese. www.priello.com

ACKNOWLEDGMENTS

Being a farmer can be a very solitary life, so I was thrilled to work on a project that involved a team effort—learning from, collaborating with, and feeding ideas to a fine group of people with a common goal. So, starting at the beginning, a big thank you to Alex, without whom I would not have met Terry McNally, who introduced me to Clare Pelino (my agent), who got me in touch with April White, who took a big messy pile of scribble that I dropped on her desk with a loud thud and organized it into a book. Clare got us in touch with Rochelle Bourgault (my editor), who works with Betsy Gammons (my photo editor), and they all prodded, coaxed, and pushed me along to help me understand the book-writing process so I was able to finish the book (almost) by my deadline. It was enjoyable, and I thank you. To the people at Quarry Books who worked to get a great-looking book together, I appreciate it and you all did a great job. Thanks also to Riccardo and Paola for the beautiful photography of my goats.

PHOTOGRAPHER CREDITS

Peter Anderson/gettyimages.com, 66
Tony Anderson/gettyimages.com, 25
© Arco Images GmbH / Alamy, 46 (bottom)
© Lena Ason / Alamy, 22 (top, right)
© Krys Bailey / Alamy, 35 (left)
Haim Behar/gettyimages.com, 5
© Peter M. Bergin/braidedbowerfarm.com, 6 (right)
© Nigel Cattlin / Alamy, 23 (top); 33 (left & right)
© Chris Clark / Alamy, 80
Simon Clay/gettyimages.com, 22 (top, left)
© Carl Costas/zumapress.com, 132
Susan Danegger / Photo Researchers, Inc., 130
Steve Debenport/gettyimages.com, 78
© Danita Delimont / Alamy, 106
Florence Delva/gettyimages.com, 62
© Alistair Dove / Alamy, 140
© Greg Balfour Evans / Alamy, 21 (top)
© Günter Fischer/agefotostock.com, 82 (left)
Fotolia.com, 26; 38; 98
Rebecca Frankeny/Bee Haven Acres/www.goatsthatfaint.com, 36 (right)
Fuse/gettyimages.com, 74
© Peter Glass/agefotostock.com, 133
© Gretchen Graham / Alamy, 145 (top)
Tim Graham/gettyimages.com, 40
Jenny Grant/Goat Justice League, 48 (middle & bottom)
© Paul Heinrich / Alamy, 50
Rachel Irving/gettyimages.com, 87
iStockphoto.com, 2; 7 (left & second, left); 12; 13; 21 (middle); 32 (left); 36 (left); 43; 48; 52; 54
Jeff Jensen/snowridgekikos.com, 35 (right)
© LOETSCHER CHLAUS / Alamy, 30
Jim McKinley/gettyimages.com, 16
Laurence Mouton/ PhotoAlto/gettyimages.com, 122
(c) Teri Myers/braidedbowerfarm.com, 91
© David Page / Alamy, 114
Ricardo Méndez Pastrana/www.ricardomendezphoto.com, 14; 21 (bottom); 22 (middle & bottom); 64; 82 (right); 86; 89 (bottom); 92; 96; 109; 112; 116; 127; 142; 146; 148; 154
© PetStockBoys / Alamy, 44
© The Photolibrary Wales / Alamy, 72
Terje Rakke/gettyimages.com, 79
© Edwin Remsberg / Alamy, 11
© Alex Rowbotham / AGRfoto / Alamy, 99
© Mark Scheuern / Alamy, 31 (left); 81
© Gorm Shackelford / Alamy, 145 (bottom)
Shutterstock.com, 6 (left, second & third, left); 7 (right & second, right); 20; 23 (middle & bottom); 42; 45; 46 (top); 47 (bottom); 48 (top); 61; 68; 75; 84; 89 (top); 100; 102;

105; 118; 119
© Mark Slack / Alamy, 47 (top)
Sodapix/photolibrary.com, 76
© inga spence / Alamy, 28
© Lynn Stone/agefotostock.com, 32 (right); 94
Washington Post/gettyimages.com, 31 (right); 90
Randy Wells/gettyimages.com, 8
© paul weston / Alamy, 29
© David Wheeldon / Alamy, 121
Courtesy of April White, 78
© WILDLIFE GmbH / Alamy, 56

INDEX